Sven Heumann

**Inverse Scattering from Chiral Media**

Sven Heumann

# Inverse Scattering from Chiral Media

## An Application of the Factorization Method

Südwestdeutscher Verlag für Hochschulschriften

**Impressum / Imprint**

Bibliografische Information der Deutschen Nationalbibliothek: Die Deutsche Nationalbibliothek verzeichnet diese Publikation in der Deutschen Nationalbibliografie; detaillierte bibliografische Daten sind im Internet über http://dnb.d-nb.de abrufbar.
Alle in diesem Buch genannten Marken und Produktnamen unterliegen warenzeichen-, marken- oder patentrechtlichem Schutz bzw. sind Warenzeichen oder eingetragene Warenzeichen der jeweiligen Inhaber. Die Wiedergabe von Marken, Produktnamen, Gebrauchsnamen, Handelsnamen, Warenbezeichnungen u.s.w. in diesem Werk berechtigt auch ohne besondere Kennzeichnung nicht zu der Annahme, dass solche Namen im Sinne der Warenzeichen- und Markenschutzgesetzgebung als frei zu betrachten wären und daher von jedermann benutzt werden dürften.

Bibliographic information published by the Deutsche Nationalbibliothek: The Deutsche Nationalbibliothek lists this publication in the Deutsche Nationalbibliografie; detailed bibliographic data are available in the Internet at http://dnb.d-nb.de.
Any brand names and product names mentioned in this book are subject to trademark, brand or patent protection and are trademarks or registered trademarks of their respective holders. The use of brand names, product names, common names, trade names, product descriptions etc. even without a particular marking in this works is in no way to be construed to mean that such names may be regarded as unrestricted in respect of trademark and brand protection legislation and could thus be used by anyone.

Coverbild / Cover image: www.ingimage.com

Verlag / Publisher:
Südwestdeutscher Verlag für Hochschulschriften
ist ein Imprint der / is a trademark of
AV Akademikerverlag GmbH & Co. KG
Heinrich-Böcking-Str. 6-8, 66121 Saarbrücken, Deutschland / Germany
Email: info@svh-verlag.de

Herstellung: siehe letzte Seite /
Printed at: see last page
**ISBN: 978-3-8381-3430-7**

Zugl. / Approved by: Karlsruhe, KIT, Diss., 2012

Copyright © 2012 AV Akademikerverlag GmbH & Co. KG
Alle Rechte vorbehalten. / All rights reserved. Saarbrücken 2012

# Preface

This work deals with several aspects of inverse scattering for chiral materials: A (chiral) body is situated in vacuum and illuminated by an electromagnetic wave. This wave is scattered. The direct scattering problem is to compute the scattered wave for a given incident wave and a given chiral object. The inverse problem is to determine the scatterer – the chiral object – from information about the scattered field.

Chiral material is optically active: the polarization is rotated when linearly–polarized plane waves pass through. It is possible to construct metamaterials which exhibit chirality for microwave frequencies.

The behavior of electromagnetic waves in chiral material is modeled by Maxwell's equations and the Drude–Born–Federov constitutive relations. We generalize the methods used for non–magnetic achiral (non–chiral) materials to treat the direct problem – existence and uniqueness – and apply the Factorization method for the reconstruction of the scatterer (inverse problem). This method delivers a necessary and sufficient condition to decide wether or not a point belongs to the scatterer.

Scattering from a bounded obstacle is studied in detail: both the direct and the inverse problem. The special case of scattering from a homogeneous chiral sphere is done analytically. Scattering by chiral cylinder is used to motivate the Factorization method for the vector Helmholtz equation. Numerical examples serve as proof of concept and illustrate the theoretical results. Finally, scattering from periodic chiral structures is another possible application of the generalized Factorization method.

This work would not exist without the support of my colleagues at the department of mathematics of the Karlsruhe Institute of Technology. First of all, I want to thank my advisor Prof. Dr. Andreas Kirsch for encouraging me to write this thesis and for valuable discussions during the recent years. I also thank PD Dr. Frank Hettlich for being co-examiner of this thesis and for his support and helpful remarks. I would like to thank all my former and present colleagues for their help and for providing a friendly working atmosphere. In particular, I am much obliged to PD Dr. Tilo Arens, Dr. Susanne Schmitt and Monika Behrens. Finally, I thank my parents for their support through my life.

# Contents

I. **Introduction**    **1**
   1. Chirality . . . . . . . . . . . . . . . . . . . . . . . . . . . . . 1
   2. Previous results and aim of the work . . . . . . . . . . . . . 4

II. **Direct transmission problem**    **9**
   1. Maxwell's equations and constitutive relations . . . . . . . . 10
   2. Variational formulation . . . . . . . . . . . . . . . . . . . . . 12
     2.1. Magnetic transmission problem . . . . . . . . . . . . . 13
     2.2. Electric transmission problem . . . . . . . . . . . . . . 17
   3. Integro–differential equation . . . . . . . . . . . . . . . . . . 22
   4. Solvability . . . . . . . . . . . . . . . . . . . . . . . . . . . . . 26
     4.1. Existence . . . . . . . . . . . . . . . . . . . . . . . . . . 26
     4.2. Uniqueness . . . . . . . . . . . . . . . . . . . . . . . . . 34

III. **Factorization Method**    **39**
   1. Far field pattern and far field operator . . . . . . . . . . . . 40
   2. Factorization of the far field operator . . . . . . . . . . . . . 51
     2.1. Modified factorization . . . . . . . . . . . . . . . . . . . 57
   3. Properties of the middle operator . . . . . . . . . . . . . . . 59
     3.1. Absorbing media . . . . . . . . . . . . . . . . . . . . . . 60
     3.2. Non–absorbing media . . . . . . . . . . . . . . . . . . . 62
   4. Localization of the scatterer . . . . . . . . . . . . . . . . . . 74

IV. **Scattering by a chiral sphere**    **83**
   1. Spherical transmission problems . . . . . . . . . . . . . . . . 84

   1.1. Spherical vector wave functions . . . . . . . . . . . . 84
   1.2. Spherical Maxwell transmission problem . . . . . . . 87
   1.3. Spherical chiral transmission problem . . . . . . . . . 90
  2. The far fiel operator . . . . . . . . . . . . . . . . . . . . . . . 94
   2.1. Series expansion of a plane wave . . . . . . . . . . . . 94
   2.2. Achiral case . . . . . . . . . . . . . . . . . . . . . . . 99
   2.3. Chiral case . . . . . . . . . . . . . . . . . . . . . . . 101

## V. Factorization Method for the vector Helmholtz case   103
  1. Motivation: Scattering by a chiral cylinder . . . . . . . . . . 104
  2. Direct transmission problem . . . . . . . . . . . . . . . . . 109
  3. Factorization Method . . . . . . . . . . . . . . . . . . . . . 117
  4. Application: Scattering by a chiral cylinder . . . . . . . . . 125
  5. Numerical Experiments . . . . . . . . . . . . . . . . . . . . 130
   5.1. Visualization of the far field pattern . . . . . . . . . . 130
   5.2. Reconstruction of the scatterer . . . . . . . . . . . . 136

## VI. Outlook: Periodic chiral media   143

## VII. Conclusions   149

## List of Symbols   151

## Index   155

## Bibliography   157

CHAPTER I

# Introduction

## 1. Chirality

The word CHIRAL comes from the greek word for hand. This describes quite well what chirality is about: handedness. In geometry, a figure is chiral if it cannot be mapped to its mirror image by rotations and translations alone. Typical examples are the human hands, snail shells and spirals and helices in general.

In chemistry, chirality usually refers to crystalline structures and molecules. The two possible configurations of a chiral molecule are called ENANTIOMERS. They are mirror images of each other. Figure I.1 shows the two enantiomers of a generic amino acid. It has a tetrahedral structure with a carbon atom at the center and four different vertices.

Material which exhibits no chirality is called ACHIRAL.

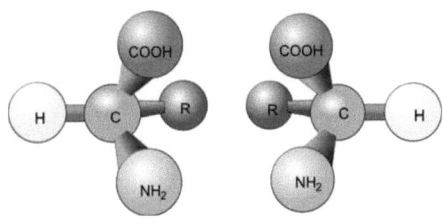

Figure I.1: The two enantiomers of a generic amino acid.

## Optical activity

Chiral material is optically active: Left– and right–circularly polarized waves propagate with different phase velocities. As a result of the right– or left– handedness of the microstructure of such materials one can observe two phenomena when linearly–polarized plane waves travel through: Circular dichroism refers to different absorption of left– and right–circularly polarized waves inside the medium. Optical rotatory dispersion describes the rotation of polarization of the transmitted wave.

The optical activity was discovered at the beginning of the 19th century. The French physicist François Arago [5] observed quartz chrystals and Jean-Baptiste Biot studied light passing through liquid solutions of tartaric acid and of sugar. Louis Pasteur found out that optical activity comes from dissymmetric arrangement of atoms in the crystalline structure or in the molecules [32]: The structures have non–superimposable mirror images and are therefore handed or chiral.

Many organic molecules exhibit chirality at optical frequencies. But chiral material must neither have a molecular origin nor is it restricted to this frequency range. Since chirality originates from the geometric property of mirror–asymmetry in the microstructure, it is possible to construct materials which are effectively chiral in sub-optical and high microwave frequencies. This may be macromolecular helical polymers embedded in non–chiral host media [38] or for example microminiature copper helices embedded in light–weight foam as in experiments studied by Lindman in 1920 [29].

According to Ammari and Nédélec [4] these artificial chiral materials attract the attention of research in the fields of scattering, waveguide propagation, antennas and microwave devices. Rojas [36] mentions the construction of conducting bodies coated with chiral materials to control their scattering properties using the additional degree of freedom given by a chirality parameter $\beta$ (to be introduced in what follows).

## Constitutive relations

The propagation of electromagnetic waves can be modeled by Maxwell's equations. In the time harmonic case with frequency $\omega$, the electric field $E$, the electric induction $D$, the magnetic field $H$ and the magnetic induction

# 1. Chirality

$B$ satisfy the equations

$$\operatorname{curl} H = -i\omega D,$$
$$\operatorname{curl} E = i\omega B,$$
$$\operatorname{div} D = 0,$$
$$\operatorname{div} B = 0.$$

Waves travelling in vacuum are adequately described by the constitutive relations

$$D = \varepsilon_0 E \quad \text{and} \quad B = \mu_0 H$$

where the constants $\varepsilon_0$ and $\mu_0$ are the permittivity and the magnetic permeability in vacuum. The constitutive relations in material media can be summarized as

$$D = \varepsilon_0 E + P \quad \text{and} \quad B = \mu_0 H + M$$

with polarization $P$ and magnetization $M$, respectively.

For non–magnetic dielectric media Lakhtakia [26] names several contributions to $P$ such as electronic, atomic or ionic and orientational polarization. They come from charge separation, displacement of atoms or ions and dipole alignments, respectively. They can be summarized as $P$ being proportional to $E$ with some real proportionality factor. In this case $D = \varepsilon E$. Lakhtakia calls $\varepsilon$ the dielectric constant. We will refer to $\varepsilon$ as absolute permittivity. If an applied electric field induces a conduction current density the ohmic losses are modeled by complex valued $\varepsilon$.

In case of magnetic media we cannot ignore magnetic dipole moments. This is modelled by $M$ being proportional to $H$. Hence, $B = \mu H$ with permeability $\mu$. Again, ohmic losses are incorporated by allowing complex values for $\mu$.

Considerations by Drude, Born and Fedorov led to the constitutive relations for isotropic chirality: the Drude–Born–Fedorov equations:

$$D = \varepsilon\bigl(E + \beta \operatorname{curl} E\bigr),$$
$$B = \mu\bigl(H + \beta \operatorname{curl} H\bigr)$$

with complex valued permittivity $\varepsilon$, magnetic permeability $\mu$ and chirality $\beta$. Here, chirality is modeled by adding a proportional dependence on $\operatorname{curl} E$ to the polarization $P$ and on $\operatorname{curl} H$ to the magnetization $M$,

| | |
|---|---|
| Condon, Charney | $D = \varepsilon_C E - \chi \partial H/\partial t$ <br> $B = \mu_C H + \chi \partial E/\partial t$ |
| Tellegen, Chambers, Unz, Krowne | $D = \varepsilon_T E + \zeta H$ <br> $B = \mu_T E - \zeta E$ |
| Post, Jaggard | $D = \varepsilon_P E + i\xi B$ <br> $B = \mu_P (H - i\xi E)$ |

Table I.1: Alternative characterizations of chirality. See Lakhtakia [26].

respectively. In his chapter about "naturally active matter" in [18] Drude argues how the curl comes into play.

Different characterizations of chirality exist. But in the time harmonic case it can be shown that they are equivalent to the one above. In Table I.1 we just cite the equations from [26] and refer to the references therein.

## 2. Previous results and aim of the work

This work deals with inverse electromagnetic scattering problems for bounded inhomogeneous chiral obstacles in 3D. The obstacle is illuminated by an incident field. We measure the scattered field, or more precisely, the far field and use this information to reconstruct the location and the shape of the scatterer. Here, we generalize the Factorization method developed by Kirsch [24] and applied to many types of inverse problems: acoustic and electromagnetic scattering for bounded obstacles, half spaces, wave guides, periodic media and electrical impedance tomography to find inclusions and for geological applications – just to name some examples. Starting point of this work is the Factorization method for Maxwell's equations and non–magnetic achiral materials – chapter 5 in [24]. The main tool – a theorem about range identity – is taken from Lechleiter [27] which is independent of interior transmission eigenvalues.

**Direct problem** In the case of homogeneous media Stratis et al. [6] use left– and right–handed fields which satisfy the achiral Maxwell equations

## 2. Previous results and aim of the work

for different wave numbers. They employ boundary integral equation methods to solve the direct (transmission) problem. The far field operator and scattering relations are studied in [7]. Rojas [36] studies the case of a chiral body attached to a perfectly electric conducting body and deduces line integrals involving the free space and chiral media Green's functions. In the case of inhomogeneous media in 3D Rojas [35] works with integral equations which are obtained with the help of the free space dyadic Green's function. Cessenat [12] proposes a variational formulation using the Calderon operator, which reduces the problem to one on a bounded domain.

We also give a variational formulation for the direct (transmission) problem motivated by a second order differential equation for one of the both fields, the electric or magnetic. We use potential solutions to particular source problems for Maxwell's equations to deduce an equivalent integro–differential equation and show existence und uniqueness.

**Inverse problem** Inverse scattering problems arise in medical imaging and non–destructive testing. There is not *the* inverse problem but many different problems can be tackled such as reconstruction of the support of the scatterer or the determination of material parameters. What kind of measurements are provided? Far field data for one single incident field or for all possible plane waves. Is the frequency fixed or is data for different frequencies available? Which a priori information is given? These questions influence uniqueness results. All inverse problems have in common to be ill–posed in the sense of Hadamard [19]: Even if one proofs existence and uniqueness the solution will not depend continuously on the data. Numerically, this fact is discouraging and challenging at the same time. Regularization schemes have to be applied and plausible a priori information has to be used for reconstruction algorithms.

Colton [13] divides the traditional approaches for the solution of inverse scattering problems into two families: non–linear optimization schemes and weak scattering approximation methods such as the Born approximation.

Bao and Peijun [8] use multi–frequency scattering data and treat the inverse medium problem with an recursive linearization on the wave number and start with the Born approximation as initial guess. Dorn et al. [17] describe an iterative method for the reconstruction of the conductivity distribution in the soil. This method can be seen as a non–linear generalization of reconstruction techniques in x–ray tomography. *Level set methods* work with an implicit representation of the unknown object and use an "evolution parameter". These methods allow to change the number of connectivity components during the reconstruction process. This has been done for homogeneous, isotropic, non–magnetic media in [34]. The disadvantages of non–linear optimization schemes are long reconstruction times and the necessity of an accurate initial guess.

Weak scattering approximation methods require a priori information whether the scatterer is impenetrable and in that case what kind of boundary condition holds. Furthermore they are not applicable in cases where the assumption of "weak scattering" is not appropriate.

A method which overcomes these difficulties is the *Linear sampling method* (LSM). The advantages are low computational effort, simple implementation and very little a priori information [13]. But it is a qualitative method in the sense that – in case of inhomogeneous scattering – one can only reconstruct the support but not the parameters of the scatterer. LSM is based on solving a linear integral equation and then using the equation's solution as an indicator function for the determination of the support of the scattering object. For more details we refer to [11] and the references therein. Another sampling method is the *No response test* proposed by Potthast and Sini [33] for the reconstruction of perfectly conducting polyhedral objects with a few incident waves. These sampling methods provide sufficient but in general not necessary conditions. The *Factorization method* developed by Kirsch [24] provides a necessary and sufficient criterion for the characteristic function of the scatterer's support.

For the inverse scattering problem in chiral media we generalize the Factorization method by combining results for scattering problems of the following kind:
$$\operatorname{curl}(\varepsilon^{-1}\operatorname{curl} u) - k^2 u = \operatorname{curl} f$$
and
$$\operatorname{curl}^2 u - k^2 \mu u = g$$
with compactly supported sources $g$ and $f$, wave number $k$, inhomogeneities $\varepsilon$ and $\mu$ and the scattered field $u$.

The chapters II and III – the direct and the inverse problem – represent the main part of this work. In chapter IV we study the scattering from spherical chiral obstacles and give series representations for the occuring

## 2. Previous results and aim of the work

fields and the far field operator. Furthermore, we study the scattering by an infinite chiral cylinder which leads to the Factorization method for the vector Helmholtz equation in chapter V. These results are mainly a corollary of the chapters II and III since the same techniques and arguments are used. Numerical experiments for the far field pattern and shape reconstruction complete this chapter. The final part consists of the chapters VI and VII where we present some ideas for the Factorization method for periodic chiral structures and give a summary and conclusions.

CHAPTER II

# Direct transmission problem

This chapter deals with the direct problem: Given an incident wave and a chiral object with known material functions compute the scattered field.

We introduce the time harmonic case for Maxwell's equations and establish the constitutive relations in vacuum and in chiral media. We discuss different forms of the equations we will deal with depending on the particular situation: Two first order equations expressing the curl of the electric field $E$ by the magnetic field $H$ and its curl and vice versa. Or expressing $\operatorname{curl} E$ and $\operatorname{curl} H$ by $E$ and $H$. In this chapter we mainly use one second order equation for $H$ or $E$, respectively.

The transmission problem is motivated and explained. We describe the setting of our problem: the chiral object, material parameters and the electromagnetic fields which appear. We use a second order differential equation for $H$ and deduce transmission conditions on the boundary of the chiral object. In this context we can call it a magnetic transmission problem. As we consider the whole space we complete the formulation by introducing the notion of a radiating solution. This handles the behaviour of the fields at infinity. All this is done classically to motivate a variational formulation. Of course, the curl–operator in a weak sense and appropriate function spaces have to be defined. Later on, we need to talk of a solution $(E, H)$ of the transmission problem. We briefly formulate an electric transmission problem and show equivalence of both, the magnetic and the electric transmission problem.

Prior to the solvability study we give an alternative formulation of our transmission problem as an integro–differential equation and show equivalence to justify the approach. Certain vector potentials are used. They are solutions to different kinds of source problems for the free space Maxwell's equations.

We use the integro–differential equation to study existence and uniqueness by interpreting it as an operator equation. The appearing operator is the sum of a coercive and a compact one. Fredholm's alternative gives us existence provided uniqueness holds. Thus, we show an uniqueness result for complex permittivity and one for smooth real valued parameters. Finally, we come back to the electric transmission problem and restate the main theorems.

## 1. Maxwell's equations and constitutive relations

In the absence of electric and magnetic currents and charges electromagnetic waves can be described in terms of the electric field $\mathcal{E}$, the electric induction $\mathcal{D}$, the magnetic field $\mathcal{H}$ and the magnetic induction $\mathcal{B}$. We treat the time harmonic case; that is,

$$\mathcal{E}(x,t) = \text{Re}\,(E(x)e^{-i\omega t}), \quad \mathcal{D}(x,t) = \text{Re}\,(D(x)e^{-i\omega t}),$$
$$\mathcal{H}(x,t) = \text{Re}\,(H(x)e^{-i\omega t}), \quad \mathcal{B}(x,t) = \text{Re}\,(B(x)e^{-i\omega t}).$$

with frequency $\omega$ and $x = (x_1, x_2, x_3)^\top \in \mathbb{R}^3$. Then the fields $E, D, H$ and $B$ satisfy Maxwell's equations in the following form

$$\text{curl}\, H = -i\omega D, \tag{2.1}$$
$$\text{curl}\, E = i\omega B, \tag{2.2}$$
$$\text{div}\, D = 0,$$
$$\text{div}\, B = 0.$$

Depending on the medium in which the waves propagate we have to add constitutive relations. We consider propagation in vacuum and chiral bodies. In vacuum we simply have

$$D = \varepsilon_0 E \quad \text{and} \quad B = \mu_0 H$$

where the constants $\varepsilon_0$ and $\mu_0$ are the permittivity and the magentic permeability in vacuum. This gives:

$$\text{curl}\, H = -i\omega\varepsilon_0 E \quad \text{and} \quad \text{curl}\, E = i\omega\mu_0 H. \tag{2.3}$$

## 1. Maxwell's equations and constitutive relations

Introduce the wave number $k := \omega\sqrt{\varepsilon_0\mu_0}$ and normalize all fields: Substitute $E$, $D$ by $E/\sqrt{\varepsilon_0}$, $D/\sqrt{\varepsilon_0}$ and $H$, $B$ by $H/\sqrt{\mu_0}$, $B/\sqrt{\mu_0}$. Then Maxwell's equations in vacuum read

$$\operatorname{curl} H = -ikE \quad \text{and} \quad \operatorname{curl} E = ikH.$$

We look at isotropic chiral media. In this case the constitutive relations (Drude–Born–Fedorov) are given by

$$D = \varepsilon\bigl(E + \beta \operatorname{curl} E\bigr) \quad \text{and} \quad B = \mu\bigl(H + \beta \operatorname{curl} H\bigr) \qquad (2.4)$$

where the relative permittivity $\varepsilon$, the relative magnetic permeability $\mu$ and the chirality $\beta$ are time independent complex valued (scalar) functions. These functions are constant for homogeneous materials. For $\beta = 0$ we recognize the achiral case.

In the next section, we use a second order differential equation for $H$ to formulate the transmission problem. Therefore we first eliminate the fields $B$ and $D$. Combining Maxwell's equations (2.1), (2.2) and the constitutive relations (2.4) we get

$$\operatorname{curl} H = -ik\varepsilon\bigl(E + \beta \operatorname{curl} E\bigr), \qquad (2.5)$$
$$\operatorname{curl} E = \phantom{-}ik\mu\bigl(H + \beta \operatorname{curl} H\bigr) \qquad (2.6)$$

For $1 - k^2\varepsilon\mu\beta^2 \neq 0$, we can express $\operatorname{curl} E$ and $\operatorname{curl} H$ by $E$ and $H$ with the aid of equations (2.5) and (2.6):

$$\operatorname{curl} H = -i \underbrace{\frac{k\varepsilon}{1 - k^2\varepsilon\mu\beta^2}}_{:=a_1} E + \underbrace{\frac{k^2\varepsilon\mu\beta}{1 - k^2\varepsilon\mu\beta^2}}_{:=a_2} H,$$
$$\operatorname{curl} E = i \underbrace{\frac{k\mu}{1 - k^2\varepsilon\mu\beta^2}}_{:=a_3} H + \underbrace{\frac{k^2\varepsilon\mu\beta}{1 - k^2\varepsilon\mu\beta^2}}_{:=a_4} E. \qquad (2.7)$$

Note that $a_2 = a_4$. Furthermore, we are able to eliminate the electric field $E$ by dividing (2.5) by $\varepsilon$ (for $\varepsilon \neq 0$) and applying the curl–Operator.

$$\tfrac{1}{\varepsilon} \operatorname{curl} H = -ikE + k^2\mu\beta H + k^2\mu\beta^2 \operatorname{curl} H,$$
$$\operatorname{curl}\bigl[\tfrac{1}{\varepsilon} \operatorname{curl} H\bigr] = k^2\mu H + k^2\mu\beta \operatorname{curl} H$$
$$\qquad\qquad + \operatorname{curl}\bigl[k^2\mu\beta H\bigr] + \operatorname{curl}\bigl[k^2\beta^2\mu \operatorname{curl} H\bigr]$$

and finally

$$\operatorname{curl}\left[\left(\tfrac{1}{\varepsilon} - k^2\mu\beta^2\right)\operatorname{curl} H\right] - k^2\left[\operatorname{curl}(\mu\beta H) + \mu\beta\operatorname{curl} H\right] - k^2\mu H = 0.$$

By the same procedure, we can eliminate the magnetic field $H$ and get a second order equation for $E$,

$$\operatorname{curl}\left[\left(\tfrac{1}{\mu} - k^2\varepsilon\beta^2\right)\operatorname{curl} E\right] - k^2\left[\operatorname{curl}(\varepsilon\beta E) + \varepsilon\beta\operatorname{curl} E\right] - k^2\varepsilon E = 0.$$

## 2. Variational formulation

We want to study the scattering of electromagnetic waves by a bounded isotropic chiral object in vacuum. When an incoming wave hits the object we can observe two phenomena. On the one hand the wave penetrates into the object. On the other hand it is scattered. How is this modeled? In the exterior, Maxwell's equations (2.3) describe the behavior of electromagnetic waves. In the interior we have the chiral version of these equations. For each of both domains we could formulate a partial differential equation. They are linked by transmission conditions on the boundary. That is why we consider a transmission problem. The direct problem consists in computing the effect – the scattered field – of an incoming wave provided the knowledge of the object – its shape and position – and the material; that is, the parameters $\varepsilon, \mu$ and $\beta$.

Motivated by the fact that we can eliminate one of the fields $E$ or $H$, we express the transmission problem with a second order partial differential equation for $E$ or $H$. We will show that both formulations lead to the same solution.

These second order equations – stated at the end of the preceding section – are symmetric in $E$ and $H$ when we interchange $\varepsilon$ and $\mu$. We formulate the two versions both classically and variationally. The classical formulation serves as motivation for the variational form.

We begin with the setting sketched in Figure II.1. A bounded chiral body characterized by the functions $\varepsilon$, $\mu$ and $\beta$ is situated in vacuum and illuminated by an electromagnetic wave $(E^i, H^i)$ causing a scattered field $(E^s, H^s)$. The total fields $E$ and $H$ are superpositions of the incident and the scattered fields,

$$E = E^i + E^s \quad \text{and} \quad H = H^i + H^s.$$

## 2. Variational formulation

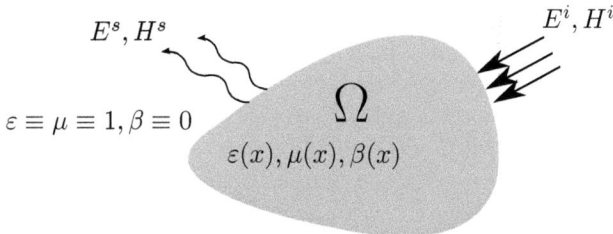

Figure II.1: Direct problem setting.

More precisely, let $\Omega \subset \mathbb{R}^3$ be a bounded domain with boundary $\Gamma \in C^2$. The functions $\varepsilon, \mu, \beta \in C^1(\mathbb{R}^3 \setminus \Gamma)$ are such that $\varepsilon \equiv 1$ $\mu \equiv 1$ and $\beta \equiv 0$ in $\mathbb{R}^3 \setminus \overline{\Omega}$. We assume that $\varepsilon \neq 0$ and $\mu \neq 0$ and introduce the contrasts $q_\mu := \mu - 1$ and $q_\varepsilon := 1 - \frac{1}{\varepsilon}$.

### 2.1. Magnetic transmission problem

We start with the total field. As we have seen in the previous section, after elimination of $E$ the governing equation for the total field $H$ is

$$\operatorname{curl}\left[\left(\tfrac{1}{\varepsilon} - k^2\mu\beta^2\right)\operatorname{curl} H\right] - k^2\left[\operatorname{curl}(\mu\beta H) + \mu\beta\operatorname{curl} H\right] - k^2\mu H = 0 \quad (2.8)$$

in $\mathbb{R}^3 \setminus \Gamma$. By the definition of the parameter functions $\varepsilon, \mu$ and $\beta$ this equation represents the free space Maxwell's equations in the exterior of $\Omega$ and the chiral equations in the interior.

As incident field $H^i$ we take an analytic solution of Maxwell's equations in vacuum,

$$\operatorname{curl}^2 H^i - k^2 H^i = 0 \quad \text{in } \mathbb{R}^3. \quad (2.9)$$

Subtraction of the equation for the incident field (2.9) from the one for the total field (2.8) yields the equation for the scattered field $H^s = H - H^i$,

$$\begin{aligned}
&\operatorname{curl}\left[\left(\tfrac{1}{\varepsilon} - k^2\mu\beta^2\right)\operatorname{curl} H^s\right] - k^2\left[\operatorname{curl}(\mu\beta H^s) + \mu\beta\operatorname{curl} H^s\right] - k^2\mu H^s \\
&= \operatorname{curl}\left[(q_\varepsilon + k^2\mu\beta^2)\operatorname{curl} H^i\right] + k^2\left[\operatorname{curl}(\mu\beta H^i) + \mu\beta\operatorname{curl} H^i\right] + k^2 q_\mu H^i
\end{aligned} \quad (2.10)$$

in $\mathbb{R}^3 \setminus \Gamma$ where the contrasts $q_\mu = \mu - 1$ and $q_\varepsilon = 1 - \frac{1}{\varepsilon}$.

The next step is to specify the transmission conditions. $\nu = \nu(x)$ denotes the unit normal vector in $x \in \Gamma = \partial\Omega$ directed to the exterior of $\Omega$. In the sequel all equations involving tangential vectors shall hold on $\Gamma$. We use the notation $F_+$ and $F_-$ for the limit from the exterior and interior, respectively, for a vector field or function $F$.

The tangential components of $E$ and $H$ are continuous on interfaces, in other words,

$$\nu \times H_+ = \nu \times H_- \quad \text{and} \quad \nu \times E_+ = \nu \times E_- \quad \text{on } \Gamma. \tag{2.11}$$

This leads to transmission conditions on the boundary $\Gamma$ of $\Omega$. The next example illustrates the determination of the transmission conditions.

**Example II.1 (Transmission conditions in achiral case).** In achiral non-magnetic media ($\beta \equiv 0$, $\mu \equiv 1$) Maxwell's equations (2.5), (2.6) read

$$\operatorname{curl} H = -ik\varepsilon E \quad \text{and} \quad \operatorname{curl} E = ikH \quad \text{in } \mathbb{R}^3 \setminus \Gamma.$$

The assumptions on $\varepsilon$ are as above: $\varepsilon \equiv 1$ in $\mathbb{R}^3 \setminus \overline{\Omega}$. We can write the continuity conditions (2.11) in terms of $H$ with the aid of Maxwell's equations. $E_- = -\frac{1}{ik\varepsilon_-} \operatorname{curl} H_-$ and $E_+ = -\frac{1}{ik} \operatorname{curl} H_+$:

$$\nu \times H_+ = \nu \times H_- \quad \text{and} \quad \nu \times \operatorname{curl} H_+ = \tfrac{1}{\varepsilon_-} \nu \times \operatorname{curl} H_-.$$

These are the transmission conditions for the total field. Those for the scattered field $H^s = H - H^i$ are obtained by subtraction. $\nu \times H_+^i = \nu \times H_-^i$ and $\nu \times \operatorname{curl} H^i = \nu \times \operatorname{curl} H_+^i = \nu \times \operatorname{curl} H_-^i$ yield

$$\nu \times H_+^s = \nu \times H_-^s$$

and

$$\tfrac{1}{\varepsilon_-} \nu \times \operatorname{curl} H_-^s - \nu \times \operatorname{curl} H_+^s = \nu \times \left(1 - \tfrac{1}{\varepsilon_-}\right) \operatorname{curl} H^i.$$

Now, we deduce the transmission conditions for the chiral case. The limits of $E$ which appear in the continuity conditions (2.11) can be expressed in terms of $H$ by the chiral equations (2.5) and (2.6). For the limit from the exterior $\beta = 0$ and $\varepsilon = 1$. Thus $E_+ = \frac{i}{k} \operatorname{curl} H_+$ and $E_- = i\left(\frac{1}{k\varepsilon} - k\mu\beta^2\right)_- \operatorname{curl} H_- - ik(\mu\beta)_- H_-$:

$$\nu \times H_+ = \nu \times H_-,$$
$$\nu \times \operatorname{curl} H_+ = \nu \times \left(\tfrac{1}{\varepsilon} - k^2\mu\beta^2\right)_- \operatorname{curl} H_- - \nu \times k^2(\mu\beta)_- H_-.$$

## 2. Variational formulation

The second condition can be rewritten

$$\nu \times \operatorname{curl} H_+ = \left(\tfrac{1}{\varepsilon} - k^2\mu\beta^2\right)_- \nu \times \operatorname{curl} H_- - k^2(\mu\beta)_- \nu \times H_-.$$

Again we get the transmission conditions in terms of the scattered field $H^s = H - H^i$ by subtraction:

$$\nu \times H^s_+ = \nu \times H^s_-$$

and

$$\left(\tfrac{1}{\varepsilon} - k^2\mu\beta^2\right)_- \nu \times \operatorname{curl} H^s_- - k^2(\mu\beta)_- \nu \times H^s_- - \nu \times \operatorname{curl} H^s_+$$
$$= \left(q_\varepsilon + k^2\mu\beta^2\right)_- \nu \times \operatorname{curl} H^i + k^2(\mu\beta)_- \nu \times H^i. \quad (2.12)$$

This looks rather complex. But in the development of a variational formulation we will recognize these expressions. They appear in the boundary integral when we integrate by parts.

### Variational formulation

Let $1/\varepsilon|_\Omega, \varepsilon|_\Omega, 1/\mu|_\Omega, \mu|_\Omega, \beta|_\Omega \in L^\infty(\Omega)$. Formally, by multiplying equation (2.10) with a test function and using integration by parts we deduce a variational formulation for $H^s$. We will show the procedure in more detail and introduce for abbreviation:

$$M^s := \left(\tfrac{1}{\varepsilon} - k^2\mu\beta^2\right) \operatorname{curl} H^s - k^2\mu\beta H^s,$$
$$m^s := k^2\mu\beta \operatorname{curl} H^s + k^2\mu H^s,$$
$$M^i := \left(q_\varepsilon + k^2\mu\beta^2\right) \operatorname{curl} H^i + k^2\mu\beta H^i,$$
$$m^i := k^2 q_\mu H^i + k^2\mu\beta \operatorname{curl} H^i.$$

Note that $M^i$ and $m^i$ vanish in $\mathbb{R}^3 \setminus \overline{\Omega}$. Then the transmission condition (2.12) is simply

$$\nu \times M^s_- - \nu \times M^s_+ = \nu \times M^i_- \quad \text{on } \Gamma$$

and the scattering equation (2.10) reads

$$\operatorname{curl} M^s - m^s = \operatorname{curl} M^i + m^i.$$

On both sides of this equation we form the scalar product with a test function $\psi \in C_0^\infty(B, \mathbb{C}^3)$ for an arbitrary ball $B \supset \overline{\Omega}$. Integration over $B$ yields

$$\iint_B \operatorname{curl} M^s \cdot \psi - m^s \cdot \psi \, \mathrm{d}x = \iint_\Omega \operatorname{curl} M^i \cdot \psi + m^i \cdot \psi \, \mathrm{d}x.$$

We split the region of integration on the left–hand side into $B \setminus \overline{\Omega}$ and $\Omega$ and apply Green's Theorem in the form

$$\iint_D \operatorname{curl} v \cdot w - v \cdot \operatorname{curl} w \, \mathrm{d}x = \int_{\partial D} \nu \cdot (v \times w) \, \mathrm{d}s = \int_{\partial D} (\nu \times v) \cdot w \, \mathrm{d}s$$

with $D = B \setminus \overline{\Omega}$ and $D = \Omega$, respectively; that is,

$$\iint_{B \setminus \overline{\Omega}} M^s \cdot \operatorname{curl} \psi - m^s \cdot \psi \, \mathrm{d}x + \iint_\Omega M^s \cdot \operatorname{curl} \psi - m^s \cdot \psi \, \mathrm{d}x$$
$$- \int_\Gamma (\nu \times M_+^s) \cdot \psi \, \mathrm{d}s + \int_\Gamma (\nu \times M_-^s) \cdot \psi \, \mathrm{d}s$$
$$= \iint_\Omega M^i \cdot \operatorname{curl} \psi + m^i \cdot \psi \, \mathrm{d}x + \int_\Gamma (\nu \times M^i) \cdot \psi \, \mathrm{d}s.$$

The boundary integrals vanish because of the transmission condition. This gives

$$\iint_{\mathbb{R}^3} M^s \cdot \operatorname{curl} \psi - m^s \cdot \psi \, \mathrm{d}x = \iint_\Omega M^i \cdot \operatorname{curl} \psi + m^i \cdot \psi \, \mathrm{d}x$$

for all $\psi$ with compact support. Plugging in the expressions for $M^s$, $m^s$, $M^i$ and $m^i$ we state the variational form of the scattering equation:

$$\iint_{\mathbb{R}^3} \left(\tfrac{1}{\varepsilon} - k^2 \mu \beta^2\right) \operatorname{curl} H^s \cdot \operatorname{curl} \psi - k^2 \mu H^s \cdot \psi \, \mathrm{d}x$$
$$- k^2 \iint_\Omega \mu \beta \left[ H^s \cdot \operatorname{curl} \psi + \operatorname{curl} H^s \cdot \psi \right] \mathrm{d}x \qquad (2.13)$$
$$= \iint_\Omega (q_\varepsilon + k^2 \mu \beta^2) \operatorname{curl} H^i \cdot \operatorname{curl} \psi + k^2 q_\mu H^i \cdot \psi \, \mathrm{d}x$$
$$+ k^2 \iint_\Omega \mu \beta \left[ H^i \cdot \operatorname{curl} \psi + \operatorname{curl} H^i \cdot \psi \right] \mathrm{d}x$$

for all $\psi$ with compact support. We specify the function spaces for $H^s, H^i$ and $\psi$ later when we formulate the weak transmission problem properly.

## 2.2. Electric transmission problem

The first section shows that the second order equations for $E$ and $H$ coincide when we interchange $\varepsilon$ and $\mu$. Analogously to the magnetic case, we can formulate the transmission problem for the electric field. We briefly give the results. Here again the incident field $E^i$ is an analytic solution to Maxwell's equations in vacuum

$$\operatorname{curl}^2 E^i - k^2 E^i = 0 \quad \text{in } \mathbb{R}^3$$

and the equation for the scattered field $E^s = E - E^i$ reads

$$\operatorname{curl}\left[\left(\tfrac{1}{\mu} - k^2\varepsilon\beta^2\right)\operatorname{curl} E^s\right] - k^2\left[\operatorname{curl}(\varepsilon\beta E^s) + \varepsilon\beta\operatorname{curl} E^s\right] - k^2\varepsilon E^s$$
$$= \operatorname{curl}\left[(p_\mu + k^2\varepsilon\beta^2)\operatorname{curl} E^i\right] + k^2\left[\operatorname{curl}(\varepsilon\beta E^i) + \varepsilon\beta\operatorname{curl} E^i\right] + k^2 p_\varepsilon E^i$$

in $\mathbb{R}^3 \setminus \Gamma$ where the contrasts $p_\varepsilon := \varepsilon - 1$ and $p_\mu := 1 - \tfrac{1}{\mu}$. The transmission conditions are

$$\nu \times E^s_+ = \nu \times E^s_-$$

and

$$\left(\tfrac{1}{\mu} - k^2\varepsilon\beta^2\right)_- \nu \times \operatorname{curl} E^s_- - k^2(\varepsilon\beta)_- \nu \times E^s_- - \nu \times \operatorname{curl} E^s_+$$
$$= (p_\mu + k^2\varepsilon\beta^2)_- \nu \times \operatorname{curl} E^i + k^2(\varepsilon\beta)_- \nu \times E^i$$

on $\Gamma$. Again $\nu$ denotes the unit normal vector on $\Gamma$ directed into the exterior of $\Omega$.

### Variational formulation

Let $1/\varepsilon|_\Omega, \varepsilon|_\Omega, 1/\mu|_\Omega, \mu|_\Omega, \beta|_\Omega \in L^\infty(\Omega)$. Again, we deduce a variational form of the scattering equation by multiplication with a test function and integrating by parts,

$$\iint_{\mathbb{R}^3} \left(\tfrac{1}{\mu} - k^2\varepsilon\beta^2\right)\operatorname{curl} E^s \cdot \operatorname{curl}\psi - k^2\varepsilon E^s \cdot \psi\, dx$$
$$-k^2 \iint_\Omega \varepsilon\beta\left[E^s \cdot \operatorname{curl}\psi + \operatorname{curl} E^s \cdot \psi\right] dx$$
$$= \iint_\Omega (p_\mu + k^2\varepsilon\beta^2)\operatorname{curl} E^i \cdot \operatorname{curl}\psi + k^2 p_\varepsilon E^i \cdot \psi\, dx \quad (2.14)$$
$$+k^2 \iint_\Omega \varepsilon\beta\left[E^i \cdot \operatorname{curl}\psi + \operatorname{curl} E^i \cdot \psi\right] dx$$

for all $\psi$ with compact support. So far, we stated two variational equations for our scattering problem. We have to specify the space in which to solve them. The first of the following definitions explains in which sense the derivatives have to be understood. We are interested in outgoing solutions. The second definition gives the notion of this property.

As in McLean [30] for any measurable subset $D \subset \mathbb{R}^3$ with strictly positive measure the function space $L^2(D)$ is defined for scalar valued functions in the usual way, equipped with the norm

$$\|u\|_{L^2(D)} := \left(\iint_D |u(x)|^2 \, \mathrm{d}x\right)^{1/2}.$$

Here and throughout this text $|\cdot|$ denotes the absolute value if the argument is scalar and $|\cdot|$ denotes the euclidean norm if the argument is a vector.

**Definition II.2 (Weak curl).** Let $D \subset \mathbb{R}^3$ a bounded domain.

(a) $L^2(D, \mathbb{C}^3) := \{v = (v_1, v_2, v_3)^\top \mid v \colon D \to \mathbb{C}^3, v_j \in L^2(B), j=1,2,3\}$

(b) A function $v \in L^2(D, \mathbb{C}^3)$ possesses a $L^2$–curl if there exists a function $w \in L^2(D, \mathbb{C}^3)$ s.t.

$$\iint_D w \cdot \psi - v \cdot \operatorname{curl} \psi \, \mathrm{d}x = 0 \qquad \text{for all } \psi \in C_0^\infty(D, \mathbb{C}^3).$$

We denote the space of these functions by $H(\operatorname{curl}, D)$.

(c) $H_{\mathrm{loc}}(\operatorname{curl}, \mathbb{R}^3) := \{v \colon \mathbb{R}^3 \to \mathbb{C}^3 \mid \forall \text{ balls } B \subset \mathbb{R}^3 : v|_B \in H(\operatorname{curl}, B)\}$

(d) The test function space

$$\{\psi \colon \mathbb{R}^3 \to \mathbb{C}^3 \mid \exists \text{ ball } B \subset \mathbb{R}^3 : \operatorname{supp} \psi \subset B, \psi|_B \in H(\operatorname{curl}, B)\}$$

is denoted by $H_{\mathrm{c}}(\operatorname{curl}, \mathbb{R}^3)$.

Notation: For functions $v$ in part (b) we use the notation $\operatorname{curl} v := w$.

**Definition II.3 (Radiating solution).** A solution $(E^s, H^s)$ to Maxwell's equations in $\mathbb{R}^3 \setminus \overline{\Omega}$ is called RADIATING if it satisfies one of the Silver–Müller radiation conditions

$$E^s(x) \times \hat{x} + H^s(x) = \mathcal{O}(|x|^{-2}) \qquad \text{as } |x| \to \infty \qquad (2.15)$$

## 2. Variational formulation

or
$$H^s(x) \times \hat{x} - E^s(x) = \mathcal{O}(|x|^{-2}) \quad \text{as } |x| \to \infty$$
uniformly with respect to $\hat{x} = x/|x|$ where $x \in \mathbb{R}^3$.

Here, $|\cdot|$ is the euclidean norm. As we will work with one of the fields we give equivalent expressions using the field and its curl.

**Proposition II.4.** *A solution $U$ to the Maxwell equations in the form*
$$\operatorname{curl}^2 U - k^2 U = 0$$
*is radiating if, and only if, $U$ satisfies one of the two conditions:*
$$\operatorname{curl} U \times \hat{x} - ikU = \mathcal{O}(|x|^{-2}) \quad \text{as } |x| \to \infty$$
*or*
$$ikU \times \hat{x} + \operatorname{curl} U = \mathcal{O}(|x|^{-2}) \quad \text{as } |x| \to \infty$$
*uniformly with respect to $\hat{x} = x/|x|$ where $x \in \mathbb{R}^3$.*

For a proof we multiply (2.15) with $-ik$ and use $\operatorname{curl} H^s = -ikE^s$. This yields the first condition of the proposition.

In the next sections we study the solvability concentrating on one of the two formulations – namely the one for $H$. But for the uniqueness result and the preliminary sections for the Factorization method we will work with both, the electric and the magnetic field and talk of a solution $(E^s, H^s)$ to the transmission problem. To justify this approach the next lemma shows: Given a solution $H^s$ to the magnetic transmission problem we can determine the corresponding electric field and vice versa.

**Lemma II.5 (Equivalence of variational formulations).** *The two variational formulations are equivalent in the following sense.*

(a) *Let $H^i$ be the incident field. If $H^s \in H_{\mathrm{loc}}(\operatorname{curl}, \mathbb{R}^3)$ is a radiating solution of (2.13) for all $\psi \in H_c(\operatorname{curl}, \mathbb{R}^3)$ then $E^s \in H_{\mathrm{loc}}(\operatorname{curl}, \mathbb{R}^3)$ defined by*

$$\begin{aligned} -ikE^s &:= \left(\tfrac{1}{\varepsilon} - k^2\mu\beta^2\right) \operatorname{curl} H^s - k^2\mu\beta H^s \\ &\quad -(q_\varepsilon + k^2\mu\beta^2) \operatorname{curl} H^i - k^2\mu\beta H^i \end{aligned} \quad (2.16)$$

*is a radiating solution of (2.14) for all $\psi \in H_c(\operatorname{curl}, \mathbb{R}^3)$ with*

$$-ikE^i := \operatorname{curl} H^i. \quad (2.17)$$

(b) Let $E^i$ be the incident field. If $E^s \in H_{\text{loc}}(\text{curl}, \mathbb{R}^3)$ is a radiating solution of (2.14) for all $\psi \in H_c(\text{curl}, \mathbb{R}^3)$ then $H^s \in H_{\text{loc}}(\text{curl}, \mathbb{R}^3)$ defined by

$$ikH^s := \left(\frac{1}{\mu} - k^2\varepsilon\beta^2\right)\text{curl}\, E^s - k^2\varepsilon\beta E^s$$
$$- (p_\mu + k^2\varepsilon\beta^2)\text{curl}\, E^i - k^2\varepsilon\beta E^i$$

is a radiating solution of (2.13) for all $\psi \in H_c(\text{curl}, \mathbb{R}^3)$ with

$$ikH^i := \text{curl}\, E^i.$$

*Proof.* (a) By the definition of $E^s$ equation (2.13) shows that $\text{curl}\, E^s$ exists locally in the weak sense and

$$-ik\,\text{curl}\, E^s = k^2\mu(\beta\,\text{curl}\, H^s + H^s) + k^2\mu\beta\,\text{curl}\, H^i + k^2 q_\mu H^i \quad (2.18)$$

in the weak sense. For $x \notin \overline{\Omega}$ we have $-ikE^s = \text{curl}\, H^s$ and

$$-ik\,\text{curl}\, E^s = k^2 H^s \Leftrightarrow \text{curl}\, E^s = ikH^s.$$

We immediately check that with $H^s$ also $E^s$ is radiating, cf. Proposition II.4. Using the total fields $H = H^s + H^i$ and $E = E^s + E^i$ equations (2.16)–(2.18) yield

$$-ik\begin{pmatrix} E \\ \text{curl}\, E \end{pmatrix} = \begin{pmatrix} \frac{1-k^2\varepsilon\mu\beta^2}{\varepsilon} & -k^2\mu\beta \\ k^2\mu\beta & k^2\mu \end{pmatrix}\begin{pmatrix} \text{curl}\, H \\ H \end{pmatrix}. \quad (2.19)$$

The determinant of the coefficient matrix is

$$\det = k^2\frac{\mu}{\varepsilon}(1 - k^2\varepsilon\mu\beta^2) + k^4\mu^2\beta^2 = k^2\frac{\mu}{\varepsilon} \in L^\infty(\Omega)$$

and the inverse matrix is given by

$$\begin{pmatrix} \varepsilon & \varepsilon\beta \\ -\varepsilon\beta & \frac{1-k^2\varepsilon\mu\beta^2}{k^2\mu} \end{pmatrix} = \frac{1}{k^2}\begin{pmatrix} k^2\varepsilon & k^2\varepsilon\beta \\ -k^2\varepsilon\beta & \frac{1-k^2\varepsilon\mu\beta^2}{\mu} \end{pmatrix}.$$

We multiply equation (2.19) with the inverse matrix:

$$\begin{pmatrix} \text{curl}\, H \\ H \end{pmatrix} = -\frac{i}{k}\begin{pmatrix} k^2\varepsilon & k^2\varepsilon\beta \\ -k^2\varepsilon\beta & \frac{1-k^2\varepsilon\mu\beta^2}{\mu} \end{pmatrix}\begin{pmatrix} E \\ \text{curl}\, E \end{pmatrix}.$$

## 2. Variational formulation

Plugging this into the definition of the weak curl,

$$\iint_{\mathbb{R}^3} H \cdot \operatorname{curl} \psi - \operatorname{curl} H \cdot \psi \, dx = 0 \quad \text{for all } \psi \in C_0^\infty(\mathbb{R}^3, \mathbb{C}^3),$$

and using again $E = E^s + E^i$ with $\operatorname{curl}^2 E^i - k^2 E^i = 0$ yields equation (2.14)
A similar computation shows part (b). □

The aim of this chapter is to show existence and uniqueness of the direct transmission problem. We finish this section with an exact formulation of that magnetic transmission problem which we want to solve. It is a generalization of the magnetic transmission problem (2.13) in two aspects:

- In the derivation of the variational formulation we see that the support of the right–hand side of the scattering equation lies in $\Omega$. We can interpret this as a source and allow more general sources $(g, h)$.

- The (real) wave number $k^2 = \omega^2 \varepsilon_0 \mu_0 > 0$ appears in several terms of equation (2.13). When we analyze the transmission problem in the following sections we have to allow complex values as well at some positions. That is the reason why we introduce the complex valued parameter $\kappa$ which replaces the wave number where it is necessary. ($\kappa$ will have the values $k$ or $ik$.)

Introduce the notation $\Pi := \{\kappa \in \mathbb{C} : \kappa \neq 0, \operatorname{Re}\kappa \geq 0, \operatorname{Im}\kappa \geq 0\}$.

**Assumption II.6** (Material parameters). *Let* $\Omega \subset \mathbb{R}^3$ *a bounded Lipschitz domain. We allow complex permittivity* $\varepsilon$ *and complex permeability* $\mu$ *but assume real chirality* $\beta$. *More precisely,* $1/\varepsilon, \mu \in L^\infty(\mathbb{R}^3, \mathbb{C})$ *and* $\beta \in L^\infty(\mathbb{R}^3, \mathbb{R})$ *such that* $\varepsilon \equiv 1$, $\mu \equiv 1$ *and* $\beta \equiv 0$ *in* $\mathbb{R}^3 \setminus \overline{\Omega}$.

**Problem 1** (Weak magnetic transmission problem). Let $k > 0$ and $\kappa \in \Pi$. Given $g, h \in L^2(\Omega, \mathbb{C}^3)$. Under the Assumption II.6 determine $v \in H_{\mathrm{loc}}(\operatorname{curl}, \mathbb{R}^3)$ such that $v$ is radiating and satisfies

$$\iint_{\mathbb{R}^3} \left(\tfrac{1}{\varepsilon} - k^2 \mu \beta^2\right) \operatorname{curl} v \cdot \operatorname{curl} \psi - \kappa^2 \mu v \cdot \psi \, dx$$

$$- \iint_\Omega \mu \beta \left[k^2 v \cdot \operatorname{curl} \psi + \kappa^2 \operatorname{curl} v \cdot \psi\right] dx \quad (2.20)$$

$$= \iint_\Omega \kappa^2 g \cdot \psi + h \cdot \operatorname{curl} \psi \, dx$$

for all $\psi \in H_c(\text{curl}, \mathbb{R}^3)$.

Additionally, we formulate the weak electric transmission problem. In this case we need not to generalize the problem for complex wave numbers.

**Assumption II.7 (Material parameters).** *Let $\Omega \subset \mathbb{R}^3$ a bounded Lipschitz domain. We allow complex permittivity $\varepsilon$ and complex permeability $\mu$ but assume real chirality $\beta$. More precisely, $\varepsilon, 1/\mu \in \underline{L^\infty}(\mathbb{R}^3, \mathbb{C})$ and $\beta \in L^\infty(\mathbb{R}^3, \mathbb{R})$ such that $\varepsilon \equiv 1$, $\mu \equiv 1$ and $\beta \equiv 0$ in $\mathbb{R}^3 \setminus \overline{\Omega}$.*

**Problem 2 (Weak electric transmission problem).** Let $k > 0$. Given sources $g, h \in L^2(\Omega, \mathbb{C}^3)$. Under the Assumption II.7 determine $v \in H_{\text{loc}}(\text{curl}, \mathbb{R}^3)$ such that $v$ is radiating and satisfies

$$\iint_{\mathbb{R}^3} \left(\tfrac{1}{\mu} - k^2 \varepsilon \beta^2\right) \text{curl}\, v \cdot \text{curl}\, \psi - k^2 \varepsilon v \cdot \psi \, dx$$

$$-k^2 \iint_\Omega \varepsilon \beta \left[v \cdot \text{curl}\, \psi + \text{curl}\, v \cdot \psi\right] dx \qquad (2.21)$$

$$= \iint_\Omega k^2 g \cdot \psi + h \cdot \text{curl}\, \psi \, dx$$

for all $\psi \in H_c(\text{curl}, \mathbb{R}^3)$.

## 3. Integro–differential equation

We give an alternative formulation for the above generalized transmission problem: an integro–differential equation (IDE). The aim is the application of Fredholm's alternative. For that reason we introduce certain vector potentials which lead to an integro–differential equation and show equivalence.

The fundamental solution to the scalar Helmholtz equation plays an important role. It will be the kernel function of our vector potentials.

**Definition II.8 (Fundamental solution).** For $\kappa \in \Pi$ the fundamental solution $\Phi_\kappa$ of the scalar Helmholtz equation in $\mathbb{R}^3$

$$\Delta u + \kappa^2 u = 0$$

## 3. Integro–differential equation

is defined by

$$\Phi_\kappa(x,y) := \frac{\exp(i\kappa|x-y|)}{4\pi|x-y|}, \qquad x \neq y.$$

The next lemma is from [21] and provides the basic vector potentials to solve Maxwell's equations.

**Lemma II.9.** *Let $\kappa \in \Pi$.*

*(a) For $f \in L^2(\Omega, \mathbb{C}^3)$ the vector field*

$$u(x) = \operatorname{curl} \iint_\Omega f(y)\Phi_\kappa(x,y)\,\mathrm{d}y, \qquad x \in \mathbb{R}^3,$$

*defines a function in $H_{\mathrm{loc}}(\operatorname{curl}, \mathbb{R}^3)$ satisfying $\operatorname{curl}^2 u - \kappa^2 u = \operatorname{curl} f$ in the variational sense; that is,*

$$\iint_{\mathbb{R}^3} \operatorname{curl} u \cdot \operatorname{curl} \psi - \kappa^2 u \cdot \psi \,\mathrm{d}x = \iint_\Omega f \cdot \operatorname{curl} \psi \,\mathrm{d}x$$

*for all $\psi \in H_c(\operatorname{curl}, \mathbb{R}^3)$. Furthermore, $u$ is radiating and the restriction $u|_\Omega$ of $u$ to $\Omega$ defines a bounded operator from $L^2(\Omega, \mathbb{C}^3)$ into $H(\operatorname{curl}, \Omega)$.*

*(b) For $f \in L^2(\Omega, \mathbb{C}^3)$ the vector field*

$$u(x) = (\kappa^2 + \nabla \operatorname{div}) \iint_\Omega f(y)\Phi_\kappa(x,y)\,\mathrm{d}y, \qquad x \in \mathbb{R}^3,$$

*defines a function in $H_{\mathrm{loc}}(\operatorname{curl}, \mathbb{R}^3)$ satisfying $\operatorname{curl}^2 u - \kappa^2 u = \kappa^2 f$ in the variational sense; that is,*

$$\iint_{\mathbb{R}^3} \operatorname{curl} u \cdot \operatorname{curl} \psi - \kappa^2 u \cdot \psi \,\mathrm{d}x = \kappa^2 \iint_\Omega f \cdot \psi \,\mathrm{d}x$$

*for all $\psi \in H_c(\operatorname{curl}, \mathbb{R}^3)$. Furthermore, $u$ is radiating and the restriction $u|_\Omega$ of $u$ to $\Omega$ defines a bounded operator from $L^2(\Omega, \mathbb{C}^3)$ into $H(\operatorname{curl}, \Omega)$.*

We develop an IDE from the above lemma. Therefore we reformulate the transmission problem (2.20) such that the variational form of the free space Maxwell's equations appears on the left–hand side.

$$\iint_{\mathbb{R}^3} \operatorname{curl} v \cdot \operatorname{curl} \psi - \kappa^2 v \cdot \psi \, dx$$

$$= \kappa^2 \iint_{\Omega} [q_\mu v + \mu\beta \operatorname{curl} v + g] \cdot \psi \, dx$$

$$+ \iint_{\Omega} [(q_\varepsilon + k^2\mu\beta^2) \operatorname{curl} v + k^2\mu\beta v + h] \cdot \operatorname{curl} \psi \, dx$$

for all $\psi \in H_c(\operatorname{curl}, \mathbb{R}^3)$. We recognize the equations from the previous lemma with $f = (q_\varepsilon + k^2\mu\beta^2)\operatorname{curl} v + k^2\mu\beta v + h$ and $f = q_\mu v + \mu\beta \operatorname{curl} v + g$, respectively. Note that $\operatorname{supp} f \subset \overline{\Omega}$ in both cases. Adapting the potentials in the previous lemma gives an IDE for $v$.

$$\begin{aligned} v(x) = (\kappa^2 + \nabla \operatorname{div}) &\iint_{\Omega} [q_\mu v + \mu\beta \operatorname{curl} v + g] \Phi_\kappa(x, \cdot) \, dy \\ + \operatorname{curl} &\iint_{\Omega} [(q_\varepsilon + k^2\mu\beta^2) \operatorname{curl} v + k^2\mu\beta v + h] \Phi_\kappa(x, \cdot) \, dy \end{aligned} \quad (2.22)$$

for $x \in \Omega$. We abbreviate $\iint_\Omega \varphi \, \Phi_\kappa(x, \cdot) \, dy := \iint_\Omega \varphi(y) \Phi_\kappa(x, y) \, dy$. We have to show that solving this integro–differential equation is equivalent to the transmission problem.

**Theorem II.10 (Equivalence).** *(a) Let $v \in H_{\mathrm{loc}}(\operatorname{curl}, \mathbb{R}^3)$ be a radiating solution of (2.20). Then $v|_\Omega \in H(\operatorname{curl}, \Omega)$ solves (2.22).*

*(b) Let $v \in H(\operatorname{curl}, \Omega)$ be a solution of (2.22). Then $v$ can be extended by the right-hand side to a radiating solution of (2.20).*

*Proof.* In this proof all partial differential equations have to be understood in the weak sense.

(a) Define $v_1$ and $v_2$ by

$$v_1(x) := (\kappa^2 + \nabla \operatorname{div}) \iint_\Omega [q_\mu v + \mu\beta \operatorname{curl} v + g] \Phi_\kappa(x, \cdot) \, dy,$$

$$v_2(x) := \operatorname{curl} \iint_\Omega [(q_\varepsilon + k^2\mu\beta^2) \operatorname{curl} v + k^2\mu\beta v + h] \Phi_\kappa(x, \cdot) \, dy$$

## 3. Integro–differential equation

for $x \in \mathbb{R}^3$. Since $v \in H_{\mathrm{loc}}(\mathrm{curl}, \mathbb{R}^3)$ is a weak solution of the transmission problem, the functions

$$q_\mu v + \mu\beta \,\mathrm{curl}\, v + g \quad \text{and} \quad (q_\varepsilon + k^2\mu\beta^2)\,\mathrm{curl}\, v + k^2\mu\beta v + h$$

are square integrable on $\Omega$ and, by Lemma II.9, $v_1$ and $v_2$ are radiating solutions in $H_{\mathrm{loc}}(\mathrm{curl}, \mathbb{R}^3)$ of

$$\mathrm{curl}^2 v_1 - \kappa^2 v_1 = \kappa^2 \left[ q_\mu v + \mu\beta\,\mathrm{curl}\, v + g \right]$$

and

$$\mathrm{curl}^2 v_2 - \kappa^2 v_2 = \mathrm{curl} \left[ (q_\varepsilon + k^2\mu\beta^2)\,\mathrm{curl}\, v + k^2\mu\beta v + h \right],$$

respectively. Therefore,

$$\begin{aligned}
&\mathrm{curl}^2 (v_1 + v_2) - \kappa^2 (v_1 + v_2) \\
&= \kappa^2 \left[ q_\mu v + \mu\beta\,\mathrm{curl}\, v + g \right] + \mathrm{curl} \left[ (q_\varepsilon + k^2\mu\beta^2)\,\mathrm{curl}\, v + k^2\mu\beta v + h \right] \\
&= \mathrm{curl}^2 v - \kappa^2 v
\end{aligned}$$

in $\mathbb{R}^3$. Both, $v_1 + v_2$ and $v$ are radiating solutions. The difference $w$, $w = v - v_1 - v_2$, is radiating and solves $\mathrm{curl}^2 w - \kappa^2 w = 0$ in $\mathbb{R}^3$. We conclude that $w \equiv 0$ and therefore $v = v_1 + v_2$ and satisfies the IDE (2.22).

(b) Let $v \in H(\mathrm{curl}, \Omega)$ be a solution of (2.22). We extend $v$ by the right-hand side to a function $\tilde{v}$ over $\mathbb{R}^3$. Then $\tilde{v}|_\Omega = v$ and $\tilde{v} \in H_{\mathrm{loc}}(\mathrm{curl}, \mathbb{R}^3)$ is a radiating solution of

$$\begin{aligned}
\mathrm{curl}^2 \tilde{v} - \kappa^2 \tilde{v} &= \kappa^2 \left[ q_\mu v + \mu\beta\,\mathrm{curl}\, v + g \right] \\
&\quad + \mathrm{curl} \left[ (q_\varepsilon + k^2\mu\beta^2)\,\mathrm{curl}\, v + k^2\mu\beta v + h \right]
\end{aligned}$$

by Lemma II.9. On $\Omega$ we have $\tilde{v} = v$ so we can write

$$\begin{aligned}
\mathrm{curl}^2 \tilde{v} - \kappa^2 \tilde{v} &= \kappa^2 \left[ q_\mu \tilde{v} + \mu\beta\,\mathrm{curl}\, \tilde{v} + g \right] \\
&\quad + \mathrm{curl} \left[ (q_\varepsilon + k^2\mu\beta^2)\,\mathrm{curl}\, \tilde{v} + k^2\mu\beta \tilde{v} + h \right]
\end{aligned}$$

whence $\tilde{v}$ is a radiating solution of (2.20). $\square$

This theorem allows us to concentrate on the IDE when studying solvability. In a similar way we can develop an IDE for the electric transmisssion problem:

**Corollary II.11.** *The equivalent integro–differential equation to the electric transmission problem (Problem 2) reads*

$$v(x) = (k^2 + \nabla \operatorname{div}) \iint_\Omega \left[ p_\varepsilon v + \varepsilon \beta \operatorname{curl} v + g \right] \Phi_k(x, \cdot) \, dy$$

$$+ \operatorname{curl} \iint_\Omega \left[ (p_\mu + k^2 \varepsilon \beta^2) \operatorname{curl} v + k^2 \varepsilon \beta v + h \right] \Phi_k(x, \cdot) \, dy$$

*for $x \in \Omega$.*

## 4. Solvability

Our goal is to solve the magnetic transmission problem (Problem 1) for $\kappa = k > 0$. In the previous section we developed an equivalent formulation, namely the integro–differential equation (2.22); that is,

$$v(x) = (k^2 + \nabla \operatorname{div}) \iint_\Omega \left[ q_\mu v + \mu \beta \operatorname{curl} v + g \right] \Phi_k(x, \cdot) \, dy$$

$$+ \operatorname{curl} \iint_\Omega \left[ (q_\varepsilon + k^2 \mu \beta^2) \operatorname{curl} v + k^2 \mu \beta v + h \right] \Phi_k(x, \cdot) \, dy.$$

Having Fredholm's alternative in mind we interpret this equation with appropriately defined operators in order to study existence and uniqueness. In the first part we reformulate the IDE as an operator equation

$$(I - A_k T_A - B_k T_B) v = f$$

with right-hand side $f$ and show that $I - A_k T_A - B_k T_B$ is a compact perturbation of an isomorphism. Here again $k$ denotes the wave number.

The second part contains two uniqueness results: one for complex material parameters and one for smooth parameter functions. On account of completeness the last subsection deals with our electric transmission problem (2.14).

### 4.1. Existence

As mentioned in the introductory part to this section we define two operators $A_\kappa$, $B_\kappa$ which are essentially the vector potentials from Lemma II.9

## 4. Solvability

and two auxiliary operators $T_A$, $T_B$ which adapt the vector potentials to our IDE. The function $f$ consists of those terms which do not depend on $v$.

**Definition II.12.** Let $k > 0$ and $\kappa \in \Pi$. Define the linear operators $A_\kappa, B_\kappa : L^2(\Omega, \mathbb{C}^3) \to H(\mathrm{curl}, \Omega)$ and $T_A, T_B : H(\mathrm{curl}, \Omega) \to L^2(\Omega, \mathbb{C}^3)$ by

$$(A_\kappa u)(x) := (\kappa^2 + \nabla \mathrm{div}) \iint_\Omega u(y)\, \Phi_\kappa(x,y)\, dy,$$

$$(B_\kappa u)(x) := \mathrm{curl} \iint_\Omega u(y)\, \Phi_\kappa(x,y)\, dy$$

for $x \in \Omega$,

$$T_A v := q_\mu v + \mu \beta \,\mathrm{curl}\, v\,, \qquad T_B v := (q_\varepsilon + k^2 \mu \beta^2)\,\mathrm{curl}\, v + k^2 \mu \beta v$$

and the function

$$f(x) := (k^2 + \nabla \mathrm{div}) \iint_\Omega g(y)\, \Phi_k(x,y)\, dy + \mathrm{curl} \iint_\Omega h(y)\, \Phi_k(x,y)\, dy$$

for $x \in \Omega$.

With these operators the above equation (2.22) simply reads

$$(I - A_k T_A - B_k T_B)\, v = f.$$

In what follows we show that this operator equation is the sum of a bounded isomorphism and a compact operator. We use the operators $A_{ik}$ and $B_{ik}$ for $k > 0$ to split up the equation:

$$(I - A_{ik} T_A - B_{ik} T_B)\, v + (A_{ik} - A_k) T_A v + (B_{ik} - B_k) T_B v = f.$$

We show that the first part on the left–hand side corresponds to a variational equation for $v$ which admits a unique solution by the Lax–Milgram lemma. The second part represents compact operators.

The next two preliminary lemmata deliver a norm equivalence result and the basic type of compact operators we are using: integral operators with weakly singular kernel functions.

**Lemma II.13.** *Let $\Omega \subset \mathbb{R}^3$ be a bounded domain and $g \in L^\infty(\Omega)$. For $v = (v_1, v_2) \in L^2(\Omega, \mathbb{C}^3)^2$ the two norms $\|\cdot\|_{L^2(\Omega, \mathbb{C}^3)^2}$ and $\|\cdot\|_g$ with*

$$\|v\|_{L^2(\Omega,\mathbb{C}^3)^2}^2 = \|v_1\|_{L^2(\Omega,\mathbb{C}^3)}^2 + \|v_2\|_{L^2(\Omega,\mathbb{C}^3)}^2 \quad \text{and } \|\cdot\|_g \text{ defined by}$$

$$\|v\|_g^2 := \|v_1 + gv_2\|_{L^2(\Omega,\mathbb{C}^3)}^2 + \|v_2\|_{L^2(\Omega,\mathbb{C}^3)}^2$$

*are equivalent.*

*Proof.* We use the inequality $a^2 + b^2 \le (a+b)^2 \le 2(a^2+b^2)$ for $a, b \ge 0$.

$$\begin{aligned}
\|v_1 + gv_2\|^2 &\le \bigl(\|v_1\| + \|g\|_{L^\infty}\|v_2\|\bigr)^2 \\
&\le 2\bigl(\|v_1\|^2 + \|g\|_{L^\infty}^2\|v_2\|^2\bigr) \\
&\le 2\max\{1, \|g\|_{L^\infty}^2\}\bigl(\|v_1\|^2 + \|v_2\|^2\bigr)
\end{aligned}$$

$$\begin{aligned}
\|v_1\|^2 + \|v_2\|^2 &\le \bigl(\|v_1\| + \|v_2\|\bigr)^2 = \bigl(\|v_1 + g\,v_2 - g\,v_2\| + \|v_2\|\bigr)^2 \\
&\le \bigl(\|v_1 + g\,v_2\| + \|g\,v_2\| + \|v_2\|\bigr)^2 \\
&\le 2\Bigl[\|v_1 + g\,v_2\|^2 + \bigl(1 + \|g\|_{L^\infty}\bigr)^2\|v_2\|^2\Bigr] \\
&\le 2\bigl(1 + \|g\|_{L^\infty}\bigr)^2\bigl(\|v_1 + g\,v_2\|^2 + \|v_2\|^2\bigr)
\end{aligned}$$

$\square$

**Definition II.14 (Volume potential).** Let $d = 2$ or $d = 3$ and let $\Omega \subset \mathbb{R}^d$ a bounded domain. Define the volume potential for a kernel function $G\colon \Omega \times \Omega \to \mathbb{C}$ by

$$V[G]\colon L^2(\Omega) \to L^2(\Omega), \qquad (V[G]\varphi)(x) := \iint_\Omega \varphi(y)\,G(x,y)\,\mathrm{d}y.$$

The kernel $G$ is called WEAKLY SINGULAR OF ORDER $\alpha$ if there exist positive constants $M$ and $\alpha \in [0, d)$ such that $G$ is continuous for all $x, y \in \Omega$ with $x \ne y$ and

$$|G(x,y)| \le M|x-y|^{-\alpha}.$$

Note that on the left–hand side $|\cdot|$ is the absolute value of a complex number and on the right–hand side $|\cdot|$ denotes the euclidian norm on $\mathbb{R}^d$.

**Lemma II.15.** *Let $d = 2$ or $d = 3$. Let $G$ be a weakly singular kernel of order $\alpha < d/2$. Then the volume potential $V[G]$ defines a compact linear operator from $L^2(\Omega)$ into $L^2(\Omega)$.*

# 4. Solvability

We sketch the proof: As in the proof of Theorem 2.21 in Kress [25] it is possible to show the compactness of the integral operator $V[G]$ from $(C(\Omega), \|\cdot\|_{L^2(\Omega)})$ into $(C(\Omega), \|\cdot\|_{L^2(\Omega)})$. In a second step one applies the following functional analysis result. For a normed space $(X, \|\cdot\|)$ denote the completion by $\tilde{X}$. Given two normed spaces $X, Y$ and a compact operator $A \colon X \to Y$. Then the unique operator $\tilde{A} \colon \tilde{X} \to \tilde{Y}$ such that $Ax = \tilde{A}x$ for $x \in X$ and $\|A\| = \|\tilde{A}\|$ is also compact. We conclude that $V[G]$ is compact from $L^2(\Omega)$ into $L^2(\Omega)$.

The conditions under which we show the main theorem are rather standard and assure the coercivness of the sesquilinear form which appears in the proof when applying the Lax–Milgram lemma.

**Assumption II.16.** *Let $k > 0$ the wave number and $\beta$ real-valued. Additionally to Assumption II.6, assume that there exist positive constants $c_1, c_2$ and $c_3 \in [0, 1)$ such that*

$$\operatorname{Re}\mu \geq c_1, \quad \operatorname{Re}\frac{1}{\varepsilon} \geq c_2 \quad \text{and} \quad k^2 \beta^2 \frac{|\varepsilon|^2 |\mu|^2}{\operatorname{Re}\varepsilon \operatorname{Re}\mu} \leq c_3 \quad \text{on } \Omega.$$

The first two conditions mean that the appearing material parameters shall be bounded away from zero. The third condition is symmetric in $\mu$ and $\varepsilon$ and depends on the wave number $k^2$ as well. It is fullfilled when $k^2 \beta^2$ is small enough or even $\beta = 0$ (achiral case).

**Theorem II.17.** *Let Assumption II.16 be fullfilled. Then:*

*(a) The operators $T_A$, $T_B$ are bounded from $H(\operatorname{curl}, \Omega)$ into $L^2(\Omega, \mathbb{C}^3)$.*

*(b) The operators $A_k - A_{ik}$ and $B_k - B_{ik}$ are compact from $L^2(\Omega, \mathbb{C}^3)$ into $H(\operatorname{curl}, \Omega)$.*

*(c) The operator $I - A_{ik}T_A - B_{ik}T_B$ is boundedly invertible in $H(\operatorname{curl}, \Omega)$.*

*Proof.* (a) $T_A v = q_\mu v + \mu\beta \operatorname{curl} v$. Straight forward estimation yields

$$\|T_A v\|_{L^2} = \|q_\mu v + \mu\beta \operatorname{curl} v\|_{L^2} \leq \|q_\mu\|_{L^\infty} \|v\|_{L^2} + \|\mu\beta\|_{L^\infty} \|\operatorname{curl} v\|_{L^2}$$
$$\leq \max\{\|q_\mu\|_{L^\infty}, \|\mu\beta\|_{L^\infty}\} \|v\|_{H(\operatorname{curl}, \Omega)}.$$

The analog holds for $T_B v = (q_\varepsilon + k^2 \mu \beta^2) \operatorname{curl} v + k^2 \mu \beta v$.

(b) We show compactness of $A_k - A_{ik}$:

$$((A_k - A_{ik})u)(x) = k^2 \iint_\Omega u(y)\, \Phi_k(x,y)\, dy + k^2 \iint_\Omega u(y)\, \Phi_{ik}(x,y)\, dy$$
$$+ \nabla \operatorname{div} \iint_\Omega u(y)\big(\Phi_k(x,y) - \Phi_{ik}(x,y)\big)\, dy.$$

The first two integrals represent three–dimensional vectors of volume potentials and define a function in $H^2(\Omega, \mathbb{C}^3)$ (cf. [15]). $H^2(\Omega, \mathbb{C}^3)$ is compactly embedded in $H^1(\Omega, \mathbb{C}^3)$. Hence, the first two integrals represent a compact operator from $L^2(\Omega, \mathbb{C}^3)$ into $H^1(\Omega, \mathbb{C}^3)$ which implies compactness from $L^2(\Omega, \mathbb{C}^3)$ into $H(\operatorname{curl}, \Omega)$.

For the third term we get

$$\nabla \operatorname{div} \iint_\Omega u\big(\Phi_k(x,\cdot) - \Phi_{ik}(x,\cdot)\big)\, dy = \iint_\Omega \nabla_x^2\big(\Phi_k(x,\cdot) - \Phi_{ik}(x,\cdot)\big) u\, dy$$
$$= \begin{pmatrix} \sum_{j=1}^3 \big(V[\frac{\partial}{\partial x_1 \partial x_j}(\Phi_k - \Phi_{ik})]u_j\big)(x) \\ \sum_{j=1}^3 \big(V[\frac{\partial}{\partial x_2 \partial x_j}(\Phi_k - \Phi_{ik})]u_j\big)(x) \\ \sum_{j=1}^3 \big(V[\frac{\partial}{\partial x_3 \partial x_j}(\Phi_k - \Phi_{ik})]u_j\big)(x) \end{pmatrix}.$$

We look at the second derivatives of the kernel in more detail. Using the expansion

$$\exp(z) = 1 + z + \frac{z^2}{2!} + \frac{z^3}{3!} + \ldots$$

we see that

$$(\Phi_k - \Phi_{ik})(x,y) = \frac{e^{ik|x-y|}}{4\pi|x-y|} - \frac{e^{-k|x-y|}}{4\pi|x-y|}$$
$$= \tfrac{1}{4\pi}\big((i+1)k - k^2|x-y| + \tfrac{1}{6}(1-i)k^3|x-y|^2 + \ldots\big)$$
$$= \tfrac{k}{4\pi}(i+1) - \tfrac{k^2}{4\pi}|x-y| + |x-y|^2 R(|x-y|)$$

where the power series $R(z) = \sum_{j=0}^\infty r_j z^j$ with constant coefficients $r_j$, $j \in \mathbb{N}_0$. We compute the gradient

$$\nabla_x\big[(\Phi_k - \Phi_{ik})(x,y)\big] = -\frac{k^2}{4\pi}\frac{x-y}{|x-y|}$$
$$+ \big[2R(|x-y|) + |x-y|R'(|x-y|)\big](x-y).$$

## 4. Solvability

The second derivative is given by the $(3 \times 3)$-matrix

$$\nabla_x^2 [(\Phi_k - \Phi_{ik})(x,y)]$$
$$= -\frac{k^2}{4\pi |x-y|} I + \frac{k^2}{4\pi |x-y|^3}(x-y)(x-y)^\top$$
$$+ \left[ 2R(|x-y|) + |x-y| R'(|x-y|) \right] I$$
$$+ \left[ 2R'(|x-y|) + R'(|x-y|) \right] \frac{1}{|x-y|}(x-y)(x-y)^\top$$
$$+ R''(|x-y|)(x-y)(x-y)^\top.$$

Hence, the second derivatives of $\Phi_k - \Phi_{ik}$ are weakly singular of order 1 which gives compactness of $A_k - A_{ik}$ as operator from $L^2$ into $L^2$. Since, $\operatorname{curl}(\nabla \operatorname{div}(\ldots)) = 0$ we also have compactness from $L^2(\Omega, \mathbb{C}^3)$ into $H(\operatorname{curl}, \Omega)$. Hence, $A_k - A_{ik}$ is compact as operator from $L^2(\Omega, \mathbb{C}^3)$ into $H(\operatorname{curl}, \Omega)$.

The compactness of $B_k - B_{ik}$ follows analogously: $B_k - B_{ik}$ represents a three–dimensional vector of volume potentials with kernel function which is weakly singular of order 0 (cf. $\nabla_x |x-y|$). Furthermore, $\operatorname{curl}(B_k - B_{ik})$ represents a three–dimensional vector of volume potentials with kernel function which is weakly singular of order 1 (compare $\nabla_x^2 |x-y|$). We conclude that $B_k - B_{ik}$ is compact from $L^2$ into $H(\operatorname{curl}, \Omega)$.

(c) For any $f \in H(\operatorname{curl}, \Omega)$ consider the equation

$$v - A_{ik} T_A v - B_{ik} T_B v = f.$$

Taking $w = v - f$ we get $w - A_{ik} T_A w - B_{ik} T_B w = A_{ik} T_A f + B_{ik} T_B f$, or explicitely,

$$w(x) = (-k^2 + \nabla \operatorname{div}) \iint_\Omega \left[ q_\mu(w+f) + \mu\beta \operatorname{curl}(w+f) \right] \Phi_{ik}(x,\cdot) \, dy$$
$$+ \operatorname{curl} \iint_\Omega \left[ (q_\varepsilon + k^2 \mu \beta^2) \operatorname{curl}(w+f) + k^2 \mu\beta(w+f) \right] \Phi_{ik}(x,\cdot) \, dy$$

for $x \in \Omega$. This equation has the form of (2.22) with $\kappa = ik$ and the functions $g$ and $h$ (in the IDE) are given by

$$g := q_\mu f + \mu\beta \operatorname{curl} f \quad \text{and} \quad h := (q_\varepsilon + k^2 \mu \beta^2) \operatorname{curl} f + k^2 \mu \beta f.$$

Hence, by Theorem II.10, $w$ can be extended to a radiating solution of the problem

$$\iint_{\mathbb{R}^3} \left[ \left(\tfrac{1}{\varepsilon} - k^2\mu\beta^2\right)\operatorname{curl} w - k^2\mu\beta w \right] \cdot \operatorname{curl}\psi + k^2\mu\left[\beta\operatorname{curl} w + w\right]\cdot\psi\,\mathrm{d}x$$
$$= \iint_{\Omega} -k^2 g\cdot\psi + h\cdot\operatorname{curl}\psi\,\mathrm{d}x \qquad (2.23)$$

for all $\psi \in H_c(\operatorname{curl},\mathbb{R}^3)$ (with the functions $g$ and $h$ as above). By definition, $w = A_{ik}T_A v + B_{ik}T_B v$. From this form and the definition of $\Phi_{ik}$ we conclude that $w$ decays exponentially as $|x|$ tends to infinity. So $w \in H(\operatorname{curl},\mathbb{R}^3)$ and the variational equation holds for all $\psi \in H(\operatorname{curl},\mathbb{R}^3)$. In order to apply the Lax–Milgram lemma we define a sesqui–linear form on $H(\operatorname{curl},\mathbb{R}^3) \times H(\operatorname{curl},\mathbb{R}^3)$ and a conjugate–linear form on $H(\operatorname{curl},\mathbb{R}^3)$.

$$a(w,\psi) := \iint_{\mathbb{R}^3} \left(\tfrac{1}{\varepsilon} - k^2\mu\beta^2\right)\operatorname{curl} w\cdot\operatorname{curl}\overline{\psi}\,\mathrm{d}x + k^2\iint_{\mathbb{R}^3}\mu w\cdot\overline{\psi}\,\mathrm{d}x$$
$$+ k^2 \iint_{\mathbb{R}^3} \mu\beta(\operatorname{curl} w\cdot\overline{\psi} - w\cdot\operatorname{curl}\overline{\psi})\,\mathrm{d}x,$$
$$b(\psi) := \iint_{\Omega} h\cdot\operatorname{curl}\overline{\psi} - k^2 g\cdot\overline{\psi}\,\mathrm{d}x.$$

$a$ and $b$ are obviously bounded.

$$\begin{aligned}
|a(w,\psi)| &\leq \left(\left\|\tfrac{1}{\varepsilon}\right\|_{L^\infty} + k^2\left\|\mu\beta^2\right\|_{L^\infty}\right)\|\operatorname{curl} w\|_{L^2}\|\operatorname{curl}\psi\|_{L^2} \\
&\quad + k^2\|\mu\beta\|_{L^\infty}\left(\|\operatorname{curl} w\|_{L^2}\|\psi\|_{L^2} + \|w\|_{L^2}\|\operatorname{curl}\psi\|_{L^2}\right) \\
&\quad + k^2\|\mu\|_{L^\infty}\|w\|_{L^2}\|\psi\|_{L^2} \\
&\leq C\left(\|\operatorname{curl} w\|_{L^2} + \|w\|_{L^2}\right)\left(\|\operatorname{curl}\psi\|_{L^2} + \|\psi\|_{L^2}\right) \\
&\leq 2C\,\|w\|_{H(\operatorname{curl}\mathbb{R}^3)}\,\|\psi\|_{H(\operatorname{curl}\mathbb{R}^3)} \\
|b(\psi)| &\leq \sqrt{2}\max\left\{\|h\|_{L^2}, k^2\|g\|_{L^2}\right\}\|\psi\|_{H(\operatorname{curl},\Omega)}
\end{aligned}$$

Here we used the inequality $x + y \leq \sqrt{2}\sqrt{x^2 + y^2}$ for $x, y \geq 0$. We show

## 4. Solvability

coercivity of $a$:

$$\begin{aligned}a(w,w) &= \iint_{\mathbb{R}^3}\left[\tfrac{1}{\varepsilon}|\operatorname{curl} w|^2 - k^2\mu\beta^2|\operatorname{curl} w|^2 + k^2\mu|w|^2\right. \\ &\qquad \left.+ k^2\mu\beta(\operatorname{curl} w\cdot\overline{w} - w\cdot\operatorname{curl}\overline{w})\right]\mathrm{d}x \\ &= \iint_{\mathbb{R}^3}\left[\tfrac{1}{\varepsilon}|\operatorname{curl} w|^2 - k^2\mu\beta^2|\operatorname{curl} w|^2 + k^2\mu|w|^2\right. \\ &\qquad \left.-2ik^2\mu\beta\operatorname{Im}(w\cdot\operatorname{curl}\overline{w})\right]\mathrm{d}x\end{aligned}$$

We take the real part of this equation and make use of the binomial $|x + iy|^2 = |x|^2 + 2\operatorname{Im}(x\overline{y}) + |y|^2$. (Recall that $\beta$ is real valued.)

$$\begin{aligned}\operatorname{Re} a(w,w) &= \\ &= \iint_{\mathbb{R}^3}\left[\operatorname{Re}\left(\tfrac{1}{\varepsilon}\right)|\operatorname{curl} w|^2 - k^2\operatorname{Re}(\mu)\beta^2|\operatorname{curl} w|^2 + \right. \\ &\qquad \left.+ k^2\operatorname{Re}(\mu)\left(|w|^2 + 2\tfrac{\operatorname{Im}\mu}{\operatorname{Re}\mu}\beta\operatorname{Im}(w\cdot\operatorname{curl}\overline{w})\right)\right]\mathrm{d}x \\ &= \iint_{\mathbb{R}^3}\left[\left(\tfrac{\operatorname{Re}\varepsilon}{|\varepsilon|^2} - k^2\beta^2\tfrac{|\mu|^2}{\operatorname{Re}\mu}\right)|\operatorname{curl} w|^2 + \right. \\ &\qquad \left.+ k^2\operatorname{Re}(\mu)\left|w + i\tfrac{\operatorname{Im}\mu}{\operatorname{Re}\mu}\beta\operatorname{curl} w\right|^2\right]\mathrm{d}x \\ &\geq c_2(1-c_3)\|\operatorname{curl} w\|_{L^2}^2 + k^2 c_1\left\|w + i\tfrac{\operatorname{Im}\mu}{\operatorname{Re}\mu}\beta\operatorname{curl} w\right\|_{L^2}^2 \\ &\geq \min(c_2(1-c_3), k^2 c_1)\left(\|\operatorname{curl} w\|_{L^2}^2 + \left\|w + i\tfrac{\operatorname{Im}\mu}{\operatorname{Re}\mu}\beta\operatorname{curl} w\right\|_{L^2}^2\right) \\ &=: \min(c_2(1-c_3), k^2 c_1)\|w\|_\beta^2\end{aligned}$$

where $\|\cdot\|_\beta$ is an equivalent norm to $\|\cdot\|_{H(\operatorname{curl},\mathbb{R}^3)}$ by Lemma II.13 with $v_1 = w$, $v_2 = \operatorname{curl} w$ and $g = i\beta\operatorname{Im}\mu/\operatorname{Re}\mu$. Now we go back to our initial equation

$$(I - A_{ik}T_A - B_{ik}T_B)v = f.$$

For given $f \in H(\operatorname{curl}, \Omega)$ we determine the (unique) solution $w$ of (2.23) and define $v := w|_\Omega + f$. Then $v - f = A_{ik}T_A v + B_{ik}T_B v$. □

With this theorem all conditions for the Fredholm alternative are satisfied. We can formulate the existence result in the next corollary.

**Corollary II.18.** *For every* $(g,h) \in L^2(\Omega, \mathbb{C}^3) \times L^2(\Omega, \mathbb{C}^3)$ *there exists a unique radiating solution* $v \in H_{\text{loc}}(\text{curl}, \mathbb{R}^3)$ *of (2.20) provided the homogeneous problem admits only the trivial solution. In that case, for any compact set* $B \supset \overline{\Omega}$ *there exists a constant* $C > 0$ *such that*

$$\|v\|_{H(\text{curl}, B)} \leq C \, \|(g,h)\|_{L^2(\Omega)^2} \quad \text{for all } (g,h) \in L^2(\Omega, \mathbb{C}^3)^2.$$

The estimate of $v$ by the source $(g,h)$ means that the solutions depends continously on the data and yields – in combination with the following uniqueness results – the well–posedness of the direct transmission problem.

We adapt the assumptions in II.16 and give the existence result for the electric transmission problem in a second corollary to the above theorem:

**Assumption II.19.** *Let* $k > 0$ *the wave number and* $\beta$ *real-valued. Additionally to Assumption II.7 assume that there exist positive constants* $c_1, c_2$ *and* $c_3 \in [0,1)$ *such that*

$$\text{Re}\,\varepsilon \geq c_1, \quad \text{Re}\,\frac{1}{\mu} \geq c_2 \quad \text{and} \quad k^2 \beta^2 \frac{|\varepsilon|^2 |\mu|^2}{\text{Re}\,\varepsilon \, \text{Re}\,\mu} \leq c_3 \quad \text{on } \Omega.$$

**Corollary II.20.** *Let Assumption II.19 be fullfilled. Then, for every source* $(g,h) \in L^2(\Omega, \mathbb{C}^3) \times L^2(\Omega, \mathbb{C}^3)$ *there exists a radiating solution* $v \in H_{\text{loc}}(\text{curl}, \mathbb{R}^3)$ *of (2.21) provided uniqueness holds. A similar estimate for* $v$ *holds.*

## 4.2. Uniqueness

Our existence results rely on the assumption that the homogeneous problem admits only the trivial solution. We give two uniqueness results.

**Theorem II.21 (Absorbing media).** *We assume, additionally to Assumption II.16, that* $\text{Im}\,\varepsilon > 0$ *and* $\text{Im}\,\mu \geq 0$ *a.e. in* $\Omega$. *Then the homogeneous magnetic transmission problem (2.20) has at most one solution (this means: uniqueness holds).*

*Proof.* Assume that $v$ is a solution of the homogeneous magnetic transmission problem namely Problem 1 for $\kappa = k$ and $v$ solves (2.20) for $g = 0$ and $h = 0$. Set $\psi = \phi \overline{v}$ in (2.20) where $\phi \in C_0^\infty(\mathbb{R}^3)$ is some mollifier with

## 4. Solvability

$\phi(x) = 1$ for $|x| \leq R$ and $\phi(x) = 0$ for $|x| \geq 2R$. $R$ is chosen such that $|x| < R$ for all $x \in \overline{\Omega}$. Then by Green's formula

$$0 = \iint_{|x|<R} [(\tfrac{1}{\varepsilon} - k^2\mu\beta^2)\operatorname{curl} v - k^2\mu\beta v] \cdot \operatorname{curl} \overline{v} - k^2\mu[\beta \operatorname{curl} v + v] \cdot \overline{v}\, dx$$
$$+ \iint_{R<|x|<2R} \operatorname{curl} v \cdot \operatorname{curl}(\phi\overline{v}) - k^2 v \cdot \phi\overline{v}\, dx \qquad (2.24)$$
$$= \iint_{|x|<R} \tfrac{1}{\varepsilon}|\operatorname{curl} v|^2 - k^2\mu|\beta \operatorname{curl} v + v|^2\, dx - \int_{|x|=R} (\operatorname{curl} v \times \nu) \cdot \overline{v}\, ds.$$

Taking the imaginary part and using $\operatorname{Im}\varepsilon > 0$ and $\operatorname{Im}\mu \geq 0$ yields

$$\operatorname{Im} \int_{|x|=R} (\operatorname{curl} v \times \nu) \cdot \overline{v}\, ds \leq 0.$$

From this we estimate

$$\int_{|x|=R} \left|\operatorname{curl} v(x) \times \tfrac{x}{|x|} - ikv(x)\right|^2 ds(x)$$
$$= \int_{|x|=R} |\operatorname{curl} v|^2 + k^2|v|^2\, ds - 2k\operatorname{Im} \int_{|x|=R} (\operatorname{curl} v \times \nu) \cdot \overline{v}\, ds$$
$$\geq \int_{|x|=R} |\operatorname{curl} v|^2 + k^2|v|^2\, ds.$$

As in the proof of Theorem 5.5 in [24] we conclude that $v$ vanishes outside of $\Omega$. Now, equation (2.24) reads

$$\iint_\Omega \tfrac{1}{\varepsilon}|\operatorname{curl} v|^2 - k^2\mu|\beta \operatorname{curl} v + v|^2\, dx = 0.$$

Taking the imaginary part yields $\operatorname{curl} v = 0$ in $\Omega$ and thus $\beta \operatorname{curl} v + v = 0$ whence $v = 0$ in $\Omega$. □

**Corollary II.22.** *Additionally to Assumption II.19, let $\operatorname{Im}\mu > 0$ and $\operatorname{Im}\varepsilon \geq 0$ a.e. in $\Omega$. Then the homoegeneous electric transmission problem (2.21) has at most one solution.*

**Theorem II.23 (Smooth media).** *Let Assumptions II.16 and II.19 be satisfied. Additionally, let $\varepsilon, \mu, \beta \in C^2(\mathbb{R}^3)$ and assume that $k^2\varepsilon\mu\beta^2 \neq 1$ in*

$\mathbb{R}^3$. *Then both, the homogeneous magnetic and the homogeneous electric transmission problem* (2.20) *and* (2.21), *respectively, have at most one solution.*

*Proof.* This proof follows Ammari and Nédélec [4] and the main argument is the unique continuation principle in Colton and Kress [15]. Assume that $v$ solves the homogeneous magnetic transmission problem for $\kappa = k$; that is, $v$ is radiating and solves (2.20) with $g = h = 0$. As in the proof of the previous theorem we conclude that $v$ vanishes outside of $\Omega$. Define the function $w$ by

$$-ikw := \left(\tfrac{1}{\varepsilon} - k^2\mu\beta^2\right)\operatorname{curl} v - k^2\mu\beta v.$$

By the weak formulation of the homogeneous problem: $w \in H_{\text{loc}}(\operatorname{curl}, \mathbb{R}^3)$. Then by Lemma II.5 $w$ is a radiating solution of the homogeneous electric transmission problem and we have

$$-ik\operatorname{curl} w = k^2\mu\beta \operatorname{curl} v + k^2\mu v.$$

From the last two equations we can deduce the system (2.7) from the first section for $H = v$ and $E = w$; that is,

$$\operatorname{curl} v = \frac{k^2\varepsilon\mu\beta}{1 - k^2\varepsilon\mu\beta^2} v - ik\frac{\varepsilon}{1 - k^2\varepsilon\mu\beta^2} w,$$

$$\operatorname{curl} w = ik\frac{\mu}{1 - k^2\varepsilon\mu\beta^2} v + \frac{k^2\varepsilon\mu\beta}{1 - k^2\varepsilon\mu\beta^2} w.$$

Now, we proceed as in Ammari and Nédélec [4]. From this system we calculate $\operatorname{curl}^2 v$, $\operatorname{curl}^2 w$, $\operatorname{div} v$ and $\operatorname{div} w$. Then we use the vector identity $\Delta = \nabla \operatorname{div} - \operatorname{curl}^2$ and apply the unique continuation principle from Colton and Kress [15] in the version of Lemma 4.15 in Monk [31]: Abbreviate

$$M = (m_{jl})_{j,l=1,2} := \frac{1}{1 - k^2\varepsilon\mu\beta^2}\begin{pmatrix} k^2\varepsilon\mu\beta & -ik\varepsilon \\ ik\mu & k^2\varepsilon\mu\beta \end{pmatrix} \in C^2(\mathbb{R}^3, \mathbb{C}^{2\times 2}).$$

Note that $\det M = -k^2\varepsilon\mu \neq 0$. Then the above equations read

$$\begin{pmatrix} \operatorname{curl} v \\ \operatorname{curl} w \end{pmatrix} = M \begin{pmatrix} v \\ w \end{pmatrix}$$

# 4. Solvability

Taking the divergence yields

$$0 = \operatorname{div}\left(M\begin{pmatrix} v \\ w \end{pmatrix}\right);$$

that is,

$$0 = \operatorname{div}(m_{11}v + m_{12}w),$$
$$0 = \operatorname{div}(m_{21}v + m_{22}w).$$

From the last two equations we conclude

$$\begin{pmatrix} \operatorname{div} v \\ \operatorname{div} w \end{pmatrix} = -\frac{1}{\det M} M^{-1}(\nabla M) \cdot \begin{pmatrix} v \\ w \end{pmatrix};$$

that is,

$$\operatorname{div} v = \frac{1}{k^2 \varepsilon \mu}(m_{22}\nabla m_{11} - m_{12}\nabla m_{21}) \cdot v$$
$$\qquad + \frac{1}{k^2 \varepsilon \mu}(m_{22}\nabla m_{12} - m_{12}\nabla m_{22}) \cdot w,$$
$$\operatorname{div} w = \frac{1}{k^2 \varepsilon \mu}(-m_{21}\nabla m_{11} + m_{11}\nabla m_{21}) \cdot v$$
$$\qquad + \frac{1}{k^2 \varepsilon \mu}(-m_{21}\nabla m_{12} + m_{22}\nabla m_{22}) \cdot w.$$

Since $v$ vanishes in the exterior of $\Omega$, also $w$ vanishes in the exterior of $\Omega$ and the traces $\nu \times v$ and $\nu \times w$ also vanish on $\partial B$ for any ball $B \supset \overline{\Omega}$. Hence, for any ball $B \supset \overline{\Omega}$: $\operatorname{curl} v \in L^2(B, \mathbb{C}^3)$, $\operatorname{div} v \in L^2(B)$ and $\nu \times v = 0$ on $\partial B$. We conclude that $v \in H^1(B, \mathbb{C}^3)$. The same holds for $w$. Compute

$$\begin{pmatrix} \operatorname{curl}^2 v \\ \operatorname{curl}^2 w \end{pmatrix} = (\nabla M) \times \begin{pmatrix} v \\ w \end{pmatrix} + M \begin{pmatrix} \operatorname{curl} v \\ \operatorname{curl} w \end{pmatrix};$$

that is,

$$\operatorname{curl} v = \nabla m_{11} \times v + \nabla m_{12} \times w + m_{11}\operatorname{curl} v + m_{12}\operatorname{curl} w,$$
$$\operatorname{curl} w = \nabla m_{21} \times v + \nabla m_{22} \times w + m_{21}\operatorname{curl} v + m_{22}\operatorname{curl} w.$$

Finally, with $\Delta = \nabla \operatorname{div} - \operatorname{curl}^2$,

$$\begin{pmatrix} \Delta v \\ \Delta w \end{pmatrix} = -\nabla \left( \frac{1}{\det M} M^{-1}(\nabla M) \begin{pmatrix} v \\ w \end{pmatrix} \right) - (\nabla M) \times \begin{pmatrix} v \\ w \end{pmatrix} + M \begin{pmatrix} \operatorname{curl} v \\ \operatorname{curl} w \end{pmatrix}$$

and $\Delta v$, $\Delta w$ exist in $L^2$. It is possible to deduce estimates of the form

$$|\Delta v_j| \leq c \sum_{l=1}^{3} |v_j| + |w_j| + |\nabla v_j| + |\nabla w_j|,$$

$$|\Delta w_j| \leq c \sum_{l=1}^{3} |v_j| + |w_j| + |\nabla v_j| + |\nabla w_j|$$

for $j = 1, 2, 3$ almost everywhere in $B$ and we can apply the unique continuation principle Lemma 4.15 in [31] which yields that $v$ (and $w$) vanish in $B \supset \overline{\Omega}$.

The same argumentation holds for the homogeneous electric transmission problem. □

This ends our study of the direct transmission problem and we are well prepared to attack the inverse problem: the localization of the scatterer.

CHAPTER III

# Factorization Method

In this chapter we solve the inverse problem: Given the solution of the direct problem or, more precisely, given the far field data, determine the scatterer $\Omega$.

The first section has preliminary character. In order to formulate the inverse problem we introduce the notion of the far field pattern which characterizes the asymptotic behavior of solutions to the transmission problem. Here the Stratton–Chu representation formulae are our main tool. They describe solutions to Maxwells equations for homogeneous media by boundary integrals. Then we can formulate the inverse problem properly and define the far field operator $\mathcal{F}$ to study it. We prove the reciprocity relation: the point of observation can be interchanged with the point observed. This allows us to give an explicite expression for the adjoint operator of $\mathcal{F}$. Furthermore we show a useful relation between $\mathcal{F}$ and $\mathcal{F}^*$.

After the definition of the far field operator and the study of some properties we deduce the eponymous factorization of $\mathcal{F}$,

$$\mathcal{F} = \mathcal{H}^* \mathcal{T} \mathcal{H}.$$

For this purpose two operators are defined. A Herglotz operator $\mathcal{H}$ maps a tangential field representing polarization vectors to an incident field. The image of its adjoint is a far field pattern. The middle operator $\mathcal{T}$ makes sure that the result of the factorization is indeed $\mathcal{F}$. The second section ends with a modified factorization for the case of non–absorbing media.

The main effort of the Factorization method is the study of the middle operator $\mathcal{T}$. Depending on the chiral material we can show different properties such as positivity and coercivity of the imaginary part or the existence of a decomposition into a coercive and a compact part. Due to the factorization we can derive spectral properties of $\mathcal{F}$. In the case of non–absorbing media we distinguish between positive and negative contrasts.

In the last section we explain the general concept of the Factorization method; that is, how to represent the characteristic function of the scatterer. A special function $\phi_z$ is defined which is contained in the range of $\mathcal{H}^*$ if, and only if, $z \in \Omega$. Finally we need a link between the range of $\mathcal{H}^*$ and $\mathcal{F}$. This can be found in abtract functional analysis theorems from [24] and [27]. We verify their assumptions by collecting the results of this chapter. This completes the Factorization method.

# 1. Far field pattern and far field operator

In the previous chapter we discussed the magnetic transmission problem. But we have seen that we can easily compute the electric field from the magnetic field which is the solution to the transmission problem. So in this section we can talk of a solution $(E^s, H^s)$ to the transmission problem. We derive the asymptotic behavior of the solution at infinity from that of the fundamental solution $\Phi_k$ with the aid of the Stratton–Chu representation formulae. Once we know the far field patterns we choose special incident fields determined by tangential fields representing polarization vectors and define the far field operator mapping the tangential field to the resulting far field pattern. Important properties of the far field operator and its adjoint are discussed then. In the proofs we use the reciprocity principle.

We begin with the asymptotic behavior of the fundamental solution and its derivatives. The formulae are taken from the proofs of Theorems 2.5 and 6.8 in Colton and Kress [15].

**Lemma III.1 (Asymptotic behavior of $\Phi_k$).** *Let $\Omega$ be a bounded domain with boundary $\Gamma$.*

*(a) The fundamental solution $\Phi_k$ has the asymptotic form*

$$\Phi_k(x,y) = \frac{e^{ik|x|}}{4\pi|x|} \left\{ e^{-ik\,\hat{x}\cdot y} + \mathcal{O}\left(\frac{1}{|x|}\right) \right\}, \qquad |x| \to \infty$$

1. Far field pattern and far field operator    41

*uniformly in all directions* $\hat{x} := x/|x|$ *for all* $y \in \Gamma$.

*(b) For any constant vector* $a \in \mathbb{C}^3$ *the derivatives of* $(a\,\Phi_k)$ *have the asymptotic form*

$$\operatorname{curl}_x a\,\Phi_k(x,y) = ik\frac{e^{ik|x|}}{4\pi|x|}\left\{e^{-ik\,\hat{x}\cdot y}(\hat{x}\times a) + \mathcal{O}\left(\frac{1}{|x|}\right)\right\},$$

$$\operatorname{curl}_x \operatorname{curl}_x a\,\Phi_k(x,y) = k^2\frac{e^{ik|x|}}{4\pi|x|}\left\{e^{-ik\,\hat{x}\cdot y}(\hat{x}\times a\times\hat{x}) + \mathcal{O}\left(\frac{1}{|x|}\right)\right\}$$

*as* $|x|\to\infty$ *uniformly for all* $y\in\Gamma$.

We continue with the well known Stratton–Chu formulae. They describe the solution of Maxwell's equations on a domain by their traces. They are taken from Colton and Kress [15]. The proofs for the weak version can be found in the book of Monk [31]. Monk also shows that the traces are well defined: Given a bounded Lipschitz domain $D$ with unit outward normal $\nu$ the mapping $v \mapsto \nu\times v|_{\partial D}$ for $v \in (C^\infty(\overline{D}))^3$ can be extended by continuity to a continuous linear map from $H(\operatorname{curl},D)$ into $H^{-1/2}(\partial D)^3$. We refer to Theorem 3.29 in [31]. We start with the Stratton–Chu formula on a bounded domain.

**Lemma III.2 (Interior Stratton–Chu).** *Assume the bounded Lipschitz domain* $\Omega$. *Let* $\nu$ *denote the unit normal vector to the boundary* $\Gamma$ *of* $\Omega$ *directed to the exterior of* $\Omega$. *Let* $E, H \in H(\operatorname{curl},\Omega)$ *be a solution to Maxwell's equations in* $\Omega$

$$\operatorname{curl} H = -ikE \quad \text{and} \quad \operatorname{curl} E = ikH. \tag{3.1}$$

*Then we have the Stratton–Chu formulae*

$$-\operatorname{curl}\int_\Gamma (\nu\times E)(y)\,\Phi_k(x,y)\,\mathrm{d}s(y)$$

$$+\frac{1}{ik}\operatorname{curl}^2\int_\Gamma (\nu\times H)(y)\,\Phi_k(x,y)\,\mathrm{d}s(y) = \begin{cases} E(x), & x\in\Omega,\\ 0, & x\in\mathbb{R}^3\setminus\overline{\Omega}, \end{cases}$$

and

$$-\operatorname{curl}\int_\Gamma (\nu \times H)(y)\,\Phi_k(x,y)\,\mathrm{d}s(y)$$
$$-\frac{1}{ik}\operatorname{curl}^2\int_\Gamma (\nu \times E)(y)\,\Phi_k(x,y)\,\mathrm{d}s(y) = \begin{cases} H(x), & x \in \Omega, \\ 0, & x \in \mathbb{R}^3 \setminus \overline{\Omega}. \end{cases}$$

Monk argues by interior regularity results that $E(x)$ and $H(x)$ – the evalutaion of $E$ and $H$ at a point $x \in \Omega$ – make sense. Furthermore, the boundary integrals have to be understood in the sense of the duality pairing between $H^{-1/2}(\Gamma)$ and $H^{1/2}(\Gamma)$. For the derivation of the far field pattern a representation formula for the exterior of bounded domains is needed:

**Lemma III.3 (Exterior Stratton–Chu).** *Assume the bounded Lipschitz domain $\Omega$ whose complement is connected. Let $\nu$ denote the unit normal vector to the boundary $\Gamma$ of $\Omega$ directed to the exterior of $\Omega$. Let $E^s, H^s \in H_{\mathrm{loc}}(\operatorname{curl}, \mathbb{R}^3 \setminus \overline{\Omega})$ be a radiating solution to Maxwell's equations in $\mathbb{R}^3 \setminus \Omega$*

$$\operatorname{curl} H^s = -ikE^s \quad \text{and} \quad \operatorname{curl} E^s = ikH^s.$$

*Then we have the Stratton–Chu formulae*

$$\operatorname{curl}\int_\Gamma (\nu \times E^s)(y)\,\Phi_k(x,y)\,\mathrm{d}s(y)$$
$$-\frac{1}{ik}\operatorname{curl}^2\int_\Gamma (\nu \times H^s)(y)\,\Phi_k(x,y)\,\mathrm{d}s(y) = \begin{cases} E^s(x), & x \in \mathbb{R}^3 \setminus \overline{\Omega}, \\ 0, & x \in \Omega, \end{cases} \quad (3.2)$$

and

$$\operatorname{curl}\int_\Gamma (\nu \times H^s)(y)\,\Phi_k(x,y)\,\mathrm{d}s(y)$$
$$+\frac{1}{ik}\operatorname{curl}^2\int_\Gamma (\nu \times E^s)(y)\,\Phi_k(x,y)\,ds(y) = \begin{cases} H^s(x), & x \in \mathbb{R}^3 \setminus \overline{\Omega}, \\ 0, & x \in \Omega. \end{cases} \quad (3.3)$$

By the formulae (3.2) and (3.3) the dependency of the fields $E^s$ and $H^s$ on $x$ is expressed by the fundamental solution. In order to determine their asymptotic behavior we only need to know their tangential traces and the

# 1. Far field pattern and far field operator

asymptotic behavior of $\Phi_k$ given in Lemma III.1. We adapt Theorem 6.8 in Colton and Kress [15] for the case of functions in $H_{\text{loc}}(\text{curl}, \mathbb{R}^3)$:

**Theorem III.4 (Far field pattern).** *Every (weak) radiating solution $E^s$, $H^s$ to the transmission problem (2.13), (2.14) for the scatterer $\Omega$ with boundary $\Gamma$ has the asymptotic form*

$$E^s(x) = \frac{e^{ik|x|}}{4\pi|x|}\left\{E^\infty(\hat{x}) + \mathcal{O}\left(\frac{1}{|x|}\right)\right\}, \quad |x| \to \infty,$$

$$H^s(x) = \frac{e^{ik|x|}}{4\pi|x|}\left\{H^\infty(\hat{x}) + \mathcal{O}\left(\frac{1}{|x|}\right)\right\}, \quad |x| \to \infty$$

*uniformly in all directions $\hat{x} = x/|x|$. The functions $E^\infty$ and $H^\infty$ defined on the unit sphere $\mathbb{S}^2$ are called* ELECTRIC *and* MAGNETIC FAR FIELD PATTERN, *respectively, and satisfy*

$$E^\infty(\hat{x}) = ik\,\hat{x} \times \int_\Gamma (\nu \times E^s)(y)\, e^{-ik\,\hat{x}\cdot y}\, ds(y)$$
$$+ ik\,\hat{x} \times \int_\Gamma (\nu \times H^s)(y)\, e^{-ik\,\hat{x}\cdot y}\, ds(y) \times \hat{x},$$

$$H^\infty(\hat{x}) = ik\,\hat{x} \times \int_\Gamma (\nu \times H^s)(y)\, e^{-ik\,\hat{x}\cdot y}\, ds(y)$$
$$- ik\,\hat{x} \times \int_\Gamma (\nu \times E^s)(y)\, e^{-ik\,\hat{x}\cdot y}\, ds(y) \times \hat{x}$$

(3.4)

*for $\hat{x} \in \mathbb{S}^2$.*

**Remark III.5.** From this theorem we observe that the far field patterns are analytic functions and tangential fields: They satisfy $E^\infty(\hat{x}) \cdot \hat{x} = 0$ and $H^\infty(\hat{x}) \cdot \hat{x} = 0$, respectively, for all $\hat{x} \in \mathbb{S}^2$. Furthermore, we easily see that, for $\hat{x} \in \mathbb{S}^2$,

$$E^\infty(\hat{x}) = H^\infty(\hat{x}) \times \hat{x} \quad \text{and} \quad H^\infty(\hat{x}) = -E^\infty(\hat{x}) \times \hat{x}.$$

In order to define the far field operator we need to specify what kind of incident fields cause the scattered and the far fields. As incident fields we consider plane waves of the form

$$\mathrm{H}^i(x;d,p) := p\,e^{ik\,d\cdot x}, \quad \mathrm{E}^i(x;d,p) = -(d \times p)e^{ik\,d\cdot x}$$

where the vectors $d \in \mathbb{S}^2$ and $p \in \mathbb{C}^3$ are the direction of incidence and polarization vector, respectively. They are chosen such that $d \cdot p = 0$ to assure that $\mathrm{H}^i$ and $\mathrm{E}^i$ are divergence free. The far field patterns $\mathrm{H}^\infty$ and $\mathrm{E}^\infty$ of the scattered fields $\mathrm{H}^s$ and $\mathrm{E}^s$ depend on $d$ and $p$ as well and we will use the notation $\mathrm{H}^\infty(\hat{x}; d, p)$ and $\mathrm{E}^\infty(\hat{x}; d, p)$, respectively.

Now we are able to formulate the inverse problem. We recall the direct one. Given an incident wave and a chiral object with known material functions compute the scattered field. If we know the scattered field we can easily compute the corresponding far field. The inverse problem is to determine the scatterer for given far field data. More precisely:

**Problem 3 (Inverse problem).** Given the wave number $k > 0$ and the data $\mathrm{H}^\infty(\hat{x}; d, p,)$ (far field pattern) for all $\hat{x}, d \in \mathbb{S}^2$ and $p \in \mathbb{C}^3$ with $p \cdot d = 0$ localize the scatterer $\Omega$.

For the study of the inverse problem we have to express it in mathematic terms; that is, we define an operator which maps a family of polarization vectors $p(d)$ characterizing the incident field to the far field pattern. Both, the family of polarization vectors and the far field pattern are tangential fields on the unit sphere.

**Definition III.6 (Far field operator).** We denote the subspace of tangential fields by $L_t^2(\mathbb{S}^2) \subset L^2(\mathbb{S}^2, \mathbb{C}^3)$; that is,

$$L_t^2(\mathbb{S}^2) := \left\{ v \in L^2(\mathbb{S}^2, \mathbb{C}^3) \ : \ v(\hat{x}) \cdot \hat{x} = 0, \hat{x} \in \mathbb{S}^2 \right\}.$$

The far field operator $\mathcal{F} \colon L_t^2(\mathbb{S}^2) \to L_t^2(\mathbb{S}^2)$ is defined by

$$(\mathcal{F}p)(\hat{x}) := \int_{\mathbb{S}^2} \mathrm{H}^\infty\bigl(\hat{x}; d, p(d)\bigr) \, \mathrm{d}s(d) \quad \text{for } \hat{x} \in \mathbb{S}^2.$$

**Remark III.7.** (a) For tangential fields $p \in L_t^2(\mathbb{S}^2)$ we have the identity

$$d \times p(d) \times d = p(d) \underbrace{(d \cdot d)}_{=1} - d \underbrace{(p(d) \cdot d)}_{=0} = p(d).$$

(b) The far field pattern $\mathrm{H}^\infty(\,\cdot\,; d, p)$ depends linearly on the polarization vector $p$. It is continuous as a function of $d$. See the proof of Theorem 6.32 in [15].

(c) Therefore, $\mathcal{F}$ is a linear integral operator with a continuous kernel. So $\mathcal{F}$ is compact. Furthermore, $\mathcal{F}p$ is the far field pattern which

1. Far field pattern and far field operator 45

corresponds to the incident field $(H_p^i, E_p^i)$ with

$$H_p^i(x) = \int_{\mathbb{S}^2} \mathrm{H}^i\bigl(x; d, p(d)\bigr) \, \mathrm{d}s(d), \quad E_p^i(x) = \int_{\mathbb{S}^2} \mathrm{E}^i\bigl(x; d, p(d)\bigr) \, \mathrm{d}s(d)$$

for $x \in \mathbb{R}^3$.

In the sequel we study the properties of the far field operator and its adjoint.

## The adjoint far field operator

The next theorem helps us to characterize the adjoint operator of $\mathcal{F}$. It describes the following reciprocity phenomenon:

We illuminate an object by a plane wave in direction $d \in \mathbb{S}^2$ with polarization $p$ and $p \cdot d = 0$. That means the incident field vectors are parallel to $p$ and their length and sign changes in direction $d$. For the measurement we choose a second pair of vectors $(\hat{x}, q) \in \mathbb{S}^2 \times \mathbb{C}^3$ with $q \cdot \hat{x} = 0$. We measure the far field which comes from direction $\hat{x}$ and just look at the polarization $q$; that is, we are interested in that part of the far field vectors which is parallel to $q$. We get the same measurements when we illuminate the object with a plane wave in direction $\hat{x}$ and polarization $q$ and measure these parts of the far field coming from direction $d$ which are parallel to $p$.

**Theorem III.8 (Reciprocity principle).** *Assume that $k^2 \varepsilon \mu \beta^2 \neq 1$. Then, for all $\hat{x}, d \in \mathbb{S}^2$ and $p, q \in \mathbb{C}^3$ with $p \cdot d = 0$ and $q \cdot \hat{x} = 0$ the reciprocity relation*

$$q \cdot \mathrm{H}^\infty(-\hat{x}; d, p) = p \cdot \mathrm{H}^\infty(-d; \hat{x}, q)$$

*holds.*

*Proof.* We abbreviate:

$$\mathrm{E}_1^i(y) := \mathrm{E}^i(y; d, p), \qquad \mathrm{E}_1^s(y) := \mathrm{E}^s(y; d, p),$$
$$\mathrm{E}_2^i(y) := \mathrm{E}^i(y; \hat{x}, q), \qquad \mathrm{E}_2^s(y) := \mathrm{E}^s(y; \hat{x}, q),$$
$$\mathrm{H}_1^i(y) := \mathrm{H}^i(y; d, p), \qquad \mathrm{H}_1^s(y) := \mathrm{H}^s(y; d, p),$$
$$\mathrm{H}_2^i(y) := \mathrm{H}^i(y; \hat{x}, q), \qquad \mathrm{H}_2^s(y) := \mathrm{H}^s(y; \hat{x}, q).$$

Application of Green's theorem (cf. Theorem 3.31 in Monk [31])

$$\iint_D \operatorname{curl} v \cdot w - v \cdot \operatorname{curl} w \, dx = \int_{\partial D} \nu \cdot (v \times w) \, ds$$

to $\mathrm{E}^i$ and $\mathrm{H}^i$ in $\Omega$ yields

$$\int_\Gamma (\nu \times \mathrm{E}_1^i) \cdot \mathrm{H}_2^i + (\nu \times \mathrm{H}_1^i) \cdot \mathrm{E}_2^i \, ds$$
$$= \iint_\Omega \operatorname{curl} \mathrm{E}_1^i \cdot \mathrm{H}_2^i - \mathrm{E}_1^i \cdot \operatorname{curl} \mathrm{H}_2^i + \operatorname{curl} \mathrm{H}_1^i \cdot \mathrm{E}_2^i - \mathrm{H}_1^i \cdot \operatorname{curl} \mathrm{E}_2^i \, dx$$
$$= \iint_\Omega ik\,\mathrm{H}_1^i \cdot \mathrm{H}_2^i + ik\,\mathrm{E}_1^i \cdot \mathrm{E}_2^i - ik\,\mathrm{E}_1^i \cdot \mathrm{E}_2^i - ik\,\mathrm{H}_1^i \cdot \mathrm{H}_2^i \, dx$$
$$= 0.$$

by Maxwell's equations (3.1). Analogously, application of Green's theorem (the traces and the boundary integral are well defined – see Monk [31]) in the exterior of $\Omega$, we obtain for $\mathrm{E}^s$ and $\mathrm{H}^s$:

$$\int_\Gamma (\nu \times \mathrm{E}_1^s) \cdot \mathrm{H}_2^s + (\nu \times \mathrm{H}_1^s) \cdot \mathrm{E}_2^s \, ds = 0.$$

Now, by the representation of the far field pattern (3.4),

$$q \cdot \mathrm{H}^\infty(-\hat{x}; d, p)$$
$$= ik \left( q \cdot (-\hat{x}) \right) \times \int_\Gamma (\nu \times \mathrm{H}_1^s)(y) \, e^{-ik(-\hat{x}) \cdot y} \, ds(y)$$
$$- ik \left( q \cdot (-\hat{x}) \right) \times \int_\Gamma (\nu \times \mathrm{E}_1^s)(y) \, e^{-ik(-\hat{x}) \cdot y} \, ds(y) \times (-\hat{x})$$
$$= -ik \int_\Gamma (\nu \times \mathrm{H}_1^s)(y) \cdot \left( (-\hat{x}) \times q \right) e^{ik\,\hat{x}\cdot y} \, ds(y)$$
$$- ik \int_\Gamma (\nu \times \mathrm{E}_1^s)(y) \cdot q \, e^{ik\,\hat{x}\cdot y} \, ds(y)$$
$$= -ik \int_\Gamma (\nu \times \mathrm{H}_1^s) \cdot \mathrm{E}_2^i + (\nu \times \mathrm{E}_1^s) \cdot \mathrm{H}_2^i \, ds.$$

Analogously:

$$p \cdot \mathrm{H}^\infty(-d; \hat{x}, q) = -ik \int_\Gamma (\nu \times \mathrm{H}_2^s) \cdot \mathrm{E}_1^i + (\nu \times \mathrm{E}_2^s) \cdot \mathrm{H}_1^i \, ds.$$

## 1. Far field pattern and far field operator 47

Thus, combination of the last four equations and application of Green's theorem yield

$$\frac{1}{ik}[q \cdot \mathrm{H}^\infty(-\hat{x}; d, p) - p \cdot \mathrm{H}^\infty(-d; \hat{x}, q)]$$
$$= \int_\Gamma (\nu \times \mathrm{H}_2) \cdot \mathrm{E}_1 + (\nu \times \mathrm{E}_2) \cdot \mathrm{H}_1 \, \mathrm{d}s$$
$$= \iint_\Omega \operatorname{curl} \mathrm{H}_2 \cdot \mathrm{E}_1 - \mathrm{H}_2 \cdot \operatorname{curl} \mathrm{E}_1 + \operatorname{curl} \mathrm{E}_2 \cdot \mathrm{H}_1 - \mathrm{E}_2 \cdot \operatorname{curl} \mathrm{H}_1 \, \mathrm{d}x.$$

As $1 - k^2\varepsilon\mu\beta^2 \neq 0$, we can express $\operatorname{curl} \mathrm{E}_j$ and $\operatorname{curl} \mathrm{H}_j$ by $\mathrm{E}_j$ and $\mathrm{H}_j$ ($j = 1, 2$) with the aid of equations (2.7).

$$\operatorname{curl} \mathrm{H}_j = -ia_1\mathrm{E}_j + a_2\mathrm{H}_j, \quad \operatorname{curl} \mathrm{E}_j = ia_3\mathrm{H}_j + a_4\mathrm{E}_j, \quad j = 1, 2.$$

We plug this into the last integral:

$$\iint_\Omega -ia_1\mathrm{E}_2 \cdot \mathrm{E}_1 + a_2\mathrm{H}_2 \cdot \mathrm{E}_1 - ia_3\mathrm{H}_2 \cdot \mathrm{H}_1 - a_4\mathrm{H}_2 \cdot \mathrm{E}_1 \, \mathrm{d}x$$
$$+ \iint_\Omega ia_3\mathrm{H}_2 \cdot \mathrm{H}_1 + a_4\mathrm{E}_2 \cdot \mathrm{H}_1 + ia_1\mathrm{E}_2 \cdot \mathrm{E}_1 - a_2\mathrm{E}_2 \cdot \mathrm{H}_1 \, \mathrm{d}x$$
$$= \iint_\Omega (a_2 - a_4)(\mathrm{H}_2 \cdot \mathrm{E}_1 - \mathrm{E}_2 \cdot \mathrm{H}_1) \, \mathrm{d}x = 0 \quad (3.5)$$

because $a_2 = a_4$. This proves the reciprocity relation. □

Due to the reciprocity principle we can give a rather explicit form for the adjoint of the far field operator.

**Corollary III.9 (Adjoint of $\mathcal{F}$).** *The adjoint* $\mathcal{F}^* : L_t^2(\mathbb{S}^2) \to L_t^2(\mathbb{S}^2)$ *of the far field operator $\mathcal{F}$ is given by*

$$(\mathcal{F}^*h)(\theta) = \int_{\mathbb{S}^2} \overline{\mathrm{H}^\infty\big(-\theta; d, \overline{h(-d)}\big)} \, \mathrm{d}s(d) \quad \text{for } \theta \in \mathbb{S}^2.$$

*Proof.*

$$\begin{aligned}
(\mathcal{F}p, h)_{L_t^2(\mathbb{S}^2)} &= \int_{\mathbb{S}^2}\int_{\mathbb{S}^2} \mathrm{H}^\infty(\theta; d, p(d)) \cdot \overline{h(\theta)} \, \mathrm{d}s(d) \, \mathrm{d}s(\theta) \\
&= \int_{\mathbb{S}^2}\int_{\mathbb{S}^2} p(d) \cdot \mathrm{H}^\infty\big(-d; -\theta, \overline{h(\theta)}\big) \, \mathrm{d}s(\theta) \, \mathrm{d}s(d) \\
&= (p, \mathcal{F}^*h)_{L_t^2(\mathbb{S}^2)}
\end{aligned}$$

where $\mathcal{F}^* h(\theta) = \int_{\mathbb{S}^2} \overline{\mathrm{H}^\infty\bigl(-\theta; d, \overline{h(-d)}\bigr)} \, \mathrm{d}s(d)$. □

The next theorem introduces the scattering operator $\mathcal{S}$ and collects some relations betweens $\mathcal{F}$, $\mathcal{F}^*$ and $\mathcal{S}$. It is adapted from Theorem 5.7 in [24]. Our proof uses the same arguments, but is formulated with the electric and magnetic field.

Recall equations (2.7)

$$\operatorname{curl} H = -i a_1 E + a_2 H \quad \text{and} \quad \operatorname{curl} E = i a_3 H + a_4 E$$

and note that $a_2 = a_4$. Introduce the matrix

$$A := \begin{pmatrix} \operatorname{Im} a_1 & -i \operatorname{Im} a_2 \\ i \operatorname{Im} a_2 & \operatorname{Im} a_3 \end{pmatrix}.$$

For the next theorem we assume that $A$ is positive semidefinite. In chapter V similar matrices and assumptions occur. Compare Proposition V.19 for possible values of $\varepsilon, \mu$ and $\beta$ such that $A$ is positive semidefinite.

**Theorem III.10 ($\mathcal{F}/\mathcal{F}^*$–Relation).** *Assume that $(A\xi) \cdot \overline{\xi} \geq 0$ on $\Omega$ for all $\xi \in \mathbb{C}^2$. Then there exists a non–negative compact self-adjoint operator $R \colon L_t^2(\mathbb{S}^2) \to L_t^2(\mathbb{S}^2)$ such that:*

*(a) The following relation holds:*

$$\mathcal{F} - \mathcal{F}^* = \frac{ik}{8\pi^2} \mathcal{F}^* \mathcal{F} + 2iR.$$

*(b) The scattering operator $\mathcal{S} := I + \frac{ik}{8\pi^2}\mathcal{F}$ is sub-unitary and has the form*

$$\mathcal{S}^* \mathcal{S} = I - \frac{k}{4\pi^2} R.$$

*$R$ vanishes for real–valued parameters. In that case $\mathcal{S}$ is unitary and $\mathcal{F}$ is normal.*

*Proof.* (a) Let $r > 0$ such that $B := B(0, r) \supset \overline{\Omega}$ and $\Omega_r = B \setminus \overline{\Omega}$. For two tangential fields $g, h \in L_t^2(\mathbb{S}^2)$ consider the incident fields

$$H_g^i = \int_{\mathbb{S}^2} \mathrm{H}^i\bigl(\,\cdot\,; d, g(d)\bigr) \, \mathrm{d}s(d), \quad H_h^i = \int_{\mathbb{S}^2} \mathrm{H}^i\bigl(\,\cdot\,; d, h(d)\bigr) \, \mathrm{d}s(d)$$

# 1. Far field pattern and far field operator

and the corresponding pairs of solutions $(E_g, H_g)$ and $(E_h, H_h)$. Then by Green's theorem

$$ik \int_{\partial B} \left[ H_g \times \overline{E_h} - E_g \times \overline{H_h} \right] \cdot \nu \, \mathrm{d}s$$

$$= ik \iint_{\Omega} \operatorname{curl} H_g \cdot \overline{E_h} - H_g \cdot \operatorname{curl} \overline{E_h} - \operatorname{curl} E_g \cdot \overline{H_h} + E_g \cdot \operatorname{curl} \overline{H_h} \, \mathrm{d}x$$

$$+ ik \iint_{\Omega_r} \operatorname{curl} H_g \cdot \overline{E_h} - H_g \cdot \operatorname{curl} \overline{E_h} - \operatorname{curl} E_g \cdot \overline{H_h} + E_g \cdot \operatorname{curl} \overline{H_h} \, \mathrm{d}x.$$

The integral over $\Omega_r$ vanishes because $(E_j, H_j)$ ($j = 1, 2$) satisfy Maxwell's equations in free space (3.1). Each total field is the sum of an incident and a scattered field. So we split the integral on the l.h.s. into four parts:

$$I_1 := ik \int_{\partial B} \left[ H_g^i \times \overline{E_h^i} - E_g^i \times \overline{H_h^i} \right] \cdot \nu \, \mathrm{d}s,$$

$$I_2 := ik \int_{\partial B} \left[ H_g^s \times \overline{E_h^s} - E_g^s \times \overline{H_h^s} \right] \cdot \nu \, \mathrm{d}s,$$

$$I_3 := ik \int_{\partial B} \left[ H_g^i \times \overline{E_h^s} - E_g^i \times \overline{H_h^s} \right] \cdot \nu \, \mathrm{d}s,$$

$$I_4 := ik \int_{\partial B} \left[ H_g^s \times \overline{E_h^i} - E_g^s \times \overline{H_h^i} \right] \cdot \nu \, \mathrm{d}s.$$

$I_1 = 0$ by Green's theorem and Maxwell's equations. By the Silver Müller radiation condition (2.15) $E^s(x) \times \hat{x} = -H^s(x) + \mathcal{O}(|x|^{-2})$ and by Theorem III.4 $H^s(x) = \frac{1}{4\pi|x|} e^{ik|x|} H^\infty(\hat{x}) + \mathcal{O}(|x|^{-2})$. Hence,

$$E^s(x) \times \hat{x} = -\frac{e^{ik|x|}}{4\pi|x|} H^\infty(\hat{x}) + \mathcal{O}(|x|^{-2})$$

and

$$ik \left( H_g^s(x) \cdot \overline{E_h^s}(x) \times \hat{x} + \overline{H_h^s}(x) \cdot E_g^s(x) \times \hat{x} \right)$$

$$= -\frac{2ik}{16\pi^2 |x|} H_g^\infty(\hat{x}) \cdot \overline{H_h^\infty}(\hat{x}) + \mathcal{O}(|x|^{-3}).$$

$$I_2 = ik \int_{\partial B} H_g^s(x) \cdot \overline{E_h^s}(x) \times \hat{x} + \overline{H_h^s}(x) \cdot E_g^s(x) \times \hat{x} \, \mathrm{d}s(x)$$

$$\longrightarrow -\frac{ik}{8\pi^2} \int_{\mathbb{S}^2} H_g^\infty \cdot \overline{H_h^\infty} \, \mathrm{d}s = -\frac{ik}{8\pi^2} (\mathcal{F}g, \mathcal{F}h)_{L^2(\mathbb{S}^2)} \quad (r \to \infty).$$

Furthermore, $I_3$ converges to $-(g, \mathcal{F}h)_{L^2}$ and $I_4$ to $(\mathcal{F}g, h)_{L^2}$. We show this for $I_3$ combining the representation formula of the far field pattern (3.4) and the explicit integral form of the incident (Herglotz) waves. Note, that the following identity holds for $p \in L_t^2(\mathbb{S}^2)$:

$$p(\theta) = \theta \times p(\theta) \times \theta, \quad \theta \in \mathbb{S}^2.$$

$$\begin{aligned} I_3 &= ik \int_{\partial B} \int_{\mathbb{S}^2} g(\theta) e^{ikx \cdot \theta} \, ds(\theta) \cdot \overline{E_h^s}(x) \times \hat{x} \, ds(x) \\ &\quad + ik \int_{\partial B} \int_{\mathbb{S}^2} \theta \times g(\theta) e^{ikx \times \theta} \, ds(\theta) \cdot \overline{H_h^s}(x) \times \hat{x} \, ds(x) \\ &= \int_{\mathbb{S}^2} g(\theta) \cdot \left[ ik\, \theta \times \int_{\partial B} \hat{x} \times \overline{H_h^s}(x) e^{ikx \cdot \theta} \, ds(x) \right] ds(\theta) \\ &\quad + \int_{\mathbb{S}^2} g(\theta) \cdot \left[ -ik\theta \times \int_{\partial B} \hat{x} \times \overline{E_h^s}(x) e^{ikx \cdot \theta} \, ds(x) \times \theta \right] ds(\theta) \\ &\stackrel{(3.4)}{=} -(g, \mathcal{F}h)_{L^2(\mathbb{S}^2)}. \end{aligned}$$

Analogously for $I_4$. One integral is left, the one over $\Omega$: Again, as in the proof of the reciprocity principle, we use equations (2.7).

$$ik \iint_\Omega \operatorname{curl} H_g \cdot \overline{E_h} - H_g \cdot \operatorname{curl} \overline{E_h} - \operatorname{curl} E_g \cdot \overline{H_h} + E_g \cdot \operatorname{curl} \overline{H_h} \, dx$$
$$= ik \iint_\Omega 2\operatorname{Im} a_1 E_g \cdot \overline{E_h} + 2\operatorname{Im} a_3 H_g \cdot \overline{H_h} + 2i \operatorname{Im} a_2 \left( H_g \cdot \overline{E_h} - E_g \cdot \overline{H_h} \right) dx.$$

Introduce the solution operator $L : L_t^2(\mathbb{S}^2) \to H(\operatorname{curl}, \Omega) \times H(\operatorname{curl}, \Omega)$, $g \mapsto (E_g|_\Omega, H_g|_\Omega)$ and recall the matrix function

$$A = \begin{pmatrix} \operatorname{Im} a_1 & -i\operatorname{Im} a_2 \\ i\operatorname{Im} a_2 & \operatorname{Im} a_3 \end{pmatrix}.$$

Then the operator $R$ is given by $Rg = kL^*(ALg)$. By assumption, $A$ is positive semidefinite, whence $R$ is non–negative. $\mathcal{F}$ is compact. Then also $\mathcal{F}^*$ is compact and therefore $R$ is compact as $2iR = \mathcal{F} - \mathcal{F}^* - \frac{ik}{8\pi^2} \mathcal{F}^* \mathcal{F}$. $A$ is self-adjoint whence $R$ is it.

(b) This assertion is the same as part (c) in Theorem 5.7 in [24] and the proof is also the same. $\square$

## 2. Factorization of the far field operator

For the remaining part of this chapter we assume that the transmission problem is uniquely solvable. Our assumptions for the following sections in detail:

**Assumption III.11.** *Let $\Omega \subset \mathbb{R}^3$ be a bounded Lipschitz domain such that the complement $\mathbb{R}^3 \setminus \overline{\Omega}$ is connected. Let $k > 0$ be the wave number and $\varepsilon, \mu \in L^\infty(\Omega, \mathbb{C})$ and $\beta \in L^\infty(\Omega, \mathbb{R})$ real–valued. We extend $\varepsilon$ and $\mu$ by one and $\beta$ by zero outside of $\Omega$. We assume:*

*(a) $1/\varepsilon \in L^\infty(\Omega, \mathbb{C})$.*

*(b) $\operatorname{Im} \varepsilon \geq 0$ and $\operatorname{Im} \mu \geq 0$ on $\Omega$.*

*(c) For all $(g, h) \in L^2(\Omega, \mathbb{C}^3) \times L^2(\Omega, \mathbb{C}^3)$ there exists a unique radiating solution of the transmission problem (2.20) for $\kappa = k$.*

We recall the main equations and notations of the transmission problem. Define an incident field $w^i$ with the aid of a tangential field $p \in L^2_t(\mathbb{S}^2)$ by

$$w^i(y) := \int_{\mathbb{S}^2} p(d) e^{ik\,d\cdot y} \mathrm{d}s(d), \quad y \in \mathbb{R}^3.$$

$w^i$ represents an analytic solution of Maxwell's equations (2.9) in vacuum. Look for a weak radiating solution $w \in H_{\mathrm{loc}}(\mathrm{curl}, \mathbb{R}^3)$ (we write $w$ instead of $w^s$) of

$$\iint_{\mathbb{R}^3} \left(\tfrac{1}{\varepsilon} - k^2 \mu \beta^2\right) \operatorname{curl} w \cdot \operatorname{curl} \psi - k^2 \mu w \cdot \psi \, \mathrm{d}x$$

$$- k^2 \iint_\Omega \mu \beta \left[w \cdot \operatorname{curl} \psi + \operatorname{curl} w \cdot \psi\right] \mathrm{d}x$$

$$= \iint_\Omega \left(q_\varepsilon + k^2 \mu \beta^2\right) \operatorname{curl} w^i \cdot \operatorname{curl} \psi + k^2 q_\mu w^i \cdot \psi \, \mathrm{d}x$$

$$+ k^2 \iint_\Omega \mu \beta \left[w^i \cdot \operatorname{curl} \psi + \operatorname{curl} w^i \cdot \psi\right] \mathrm{d}x$$

for all $\psi \in H_c(\mathrm{curl}, \mathbb{R}^3)$. The tangential field $p$ determines the incident field $w^i$. The far field operator $\mathcal{F}$ maps $p$ to the far field $w^\infty$ of the scattered field $w$ caused by $w^i$,

$$\mathcal{F}p = w^\infty.$$

In the following, we deduce a factorization of $\mathcal{F}$; that is, $\mathcal{F} = \mathcal{H}^*\mathcal{T}\mathcal{H}$. We begin with the operator $\mathcal{H}$ which is some kind of Herglotz operator and maps a tangential field on $\mathbb{S}^2$ to an $L^2$ function in $\Omega$.
Notation: $L^2(\Omega, \mathbb{C}^3)^2 := L^2(\Omega, \mathbb{C}^3) \times L^2(\Omega, \mathbb{C}^3)$

**Definition III.12 (Herglotz operator).** Let $k > 0$. Define two linear operators $\mathcal{H}_j \colon L^2_t(\mathbb{S}^2) \to L^2(\Omega, \mathbb{C}^3)$ $(j = 1, 2)$ by

$$(\mathcal{H}_1 p)(y) := \int_{\mathbb{S}^2} p(d) e^{ik\, d \cdot y}\, ds(d) \quad \text{and} \quad \mathcal{H}_2 p := \operatorname{curl}\left[\mathcal{H}_1 p\right]$$

for $p \in L^2_t(\mathbb{S}^2)$ and $y \in \Omega$. Then the HERGLOTZ OPERATOR is defined by $\mathcal{H} \colon L^2_t(\mathbb{S}^2) \to L^2(\Omega, \mathbb{C}^3)^2$,

$$\mathcal{H}p = (\mathcal{H}_1 p, \mathcal{H}_2 p)^\top.$$

**Remark III.13.** $\mathcal{H}$ is a bounded operator and injective: $\mathcal{H}p = 0$ implies $\int_{\mathbb{S}^2} p(d) e^{ikd\cdot x}\, ds(d) = 0$ and this implies $p \equiv 0$, see Colton and Kress [15].

We introduce the DATA–TO–PATTERN OPERATOR $G$ which maps the "source" $f = (f_1, f_2)$ to the far field pattern,

$$Gf := v^\infty.$$

where $v^\infty$ is the far field pattern of the radiating solution of

$$\iint_{\mathbb{R}^3} \left[(\tfrac{1}{\varepsilon} - k^2\mu\beta^2)\operatorname{curl} v - k^2\mu\beta v\right] \cdot \operatorname{curl}\psi - k^2\left[\mu\beta \operatorname{curl} v + \mu v\right]\cdot \psi\, dx$$
$$= \iint_\Omega k^2\left[q_\mu f_1 + \mu\beta f_2\right] \cdot \psi + \left[(q_\varepsilon + k^2\mu\beta^2)f_2 + k^2\mu\beta f_1\right] \cdot \operatorname{curl}\psi\, dx$$
(3.6)

for all $\psi \in H_c(\operatorname{curl}, \mathbb{R}^3)$. The auxiliary scattering equation (3.6) is just designed such that $\mathcal{F}p = G\mathcal{H}p$. If we have a look at our goal, namely the product $\mathcal{F} = \mathcal{H}^*\mathcal{T}\mathcal{H}$, we need the adjoint operator of $\mathcal{H}$ and show that $G = \mathcal{H}^*\mathcal{T}$ for an operator $\mathcal{T}$ to be specified. In the next proposition we compute the adjoint operators of $\mathcal{H}_1$ and $\mathcal{H}_2$.

## 2. Factorization of the far field operator

**Proposition III.14 (Adjoint Herglotz operator).** *The adjoint operators* $\mathcal{H}_1^*, \mathcal{H}_2^* \colon L^2(\Omega, \mathbb{C}^3) \to L_t^2(\mathbb{S}^2)$ *of* $\mathcal{H}_1$ *and* $\mathcal{H}_2$ *are given by*

$$(\mathcal{H}_1^*\varphi)(d) = d \times \iint_\Omega \varphi(y)\, e^{-ik\, d\cdot y}\mathrm{d}y \times d$$

$$(\mathcal{H}_2^*\varphi)(d) = ik\, d \times \iint_\Omega \varphi(y)\, e^{-ik\, d\cdot y}\mathrm{d}y$$

*for* $d \in \mathbb{S}^2$. *Thus the adjoint* $\mathcal{H}^* \colon L^2(\Omega, \mathbb{C}^3)^2 \to L_t^2(\mathbb{S}^2)$ *is given by*

$$\mathcal{H}^*\varphi = \mathcal{H}_1^*\varphi_1 + \mathcal{H}_2^*\varphi_2$$

*for* $\varphi = (\varphi_1, \varphi_2) \in L^2(\Omega, \mathbb{C}^3)^2$.

*Proof.* We start with $\mathcal{H}_1$. By Remark III.7 $p(d) = d \times p(d) \times d$ and therefore

$$\begin{aligned}
(\mathcal{H}_1 p, \varphi)_{L^2(\Omega,\mathbb{C}^3)} &= \iint_\Omega \int_{\mathbb{S}^2} \bigl(d \times p(d) \times d\bigr) e^{ik\, d\cdot y}\mathrm{d}s(d) \cdot \overline{\varphi(y)}\, \mathrm{d}y \\
&= \int_{\mathbb{S}^2} \iint_\Omega \bigl(d \times p(d) \times d\bigr) \cdot \overline{\varphi(y)}\, e^{ik\, d\cdot y}\mathrm{d}y\, \mathrm{d}s(d) \\
&= \int_{\mathbb{S}^2} p(d) \cdot \overline{d \times \iint_\Omega \varphi(y) e^{-ik\, d\cdot y}\, \mathrm{d}y \times d}\, \mathrm{d}s(d) \\
&= (p, \mathcal{H}_1^*\varphi)_{L_t^2(\mathbb{S}^2)}
\end{aligned}$$

where $(\mathcal{H}_1^*\varphi)(d) = d \times \iint_\Omega \varphi(y) e^{-ik\, d\cdot y}\, \mathrm{d}y \times d$.

For $\mathcal{H}_2$ we compute $\operatorname{curl} \int_{\mathbb{S}^2} p(d) e^{ik\, d\cdot y}\, \mathrm{d}s(d) = ik\, d \times \int_{\mathbb{S}^2} p(d) e^{ik\, d\cdot y}\, \mathrm{d}s(d)$ and

$$\begin{aligned}
(\mathcal{H}_2 p, \varphi)_{L^2(\Omega,\mathbb{C}^3)} &= \iint_\Omega \operatorname{curl} \int_{\mathbb{S}^2} p(d)\, e^{ik\, d\cdot y}\, \mathrm{d}s(d) \cdot \overline{\varphi(y)}\, \mathrm{d}y \\
&= \int_{\mathbb{S}^2} \iint_\Omega ik\, \bigl(d \times p(d)\bigr) \cdot \overline{\varphi(y)}\, e^{ik\, d\cdot y}\, \mathrm{d}y\, \mathrm{d}s(d) \\
&= \int_{\mathbb{S}^2} p(d) \cdot \overline{(ik\, d) \times \iint_\Omega \varphi(y)\, e^{-ik\, d\cdot y}\, \mathrm{d}y}\, \mathrm{d}s(d) \\
&= (p, \mathcal{H}_2^*\varphi)_{L_t^2(\mathbb{S}^2)}
\end{aligned}$$

where $(\mathcal{H}_2^*\varphi)(d) = ikd \times \int_\Omega \varphi(y) e^{-ikd\cdot y} \, dy$. Finally,

$$\begin{aligned}(\mathcal{H}p, \varphi)_{L^2(\Omega,\mathbb{C}^3)^2} &= (\mathcal{H}_1 p, \varphi_1)_{L^2(\Omega,\mathbb{C}^3)} + (\mathcal{H}_2 p, \varphi_2)_{L^2(\Omega,\mathbb{C}^3)} \\ &= (p, \mathcal{H}_1^*\varphi_1)_{L_t^2(\mathbb{S}^2)} + (p, \mathcal{H}_2^*\varphi_2)_{L_t^2(\mathbb{S}^2)}.\end{aligned}$$

Thus, $\mathcal{H}^*(\varphi_1, \varphi_2) = \mathcal{H}_1^*\varphi_1 + \mathcal{H}_2^*\varphi_2$. □

What is the image of $\mathcal{H}^*$? It should be a far field pattern. We extend Lemma II.9 and calculate the far field patterns related to the solutions in form of potentials.

**Proposition III.15.** *Let $k > 0$. For $f_1, f_2 \in L^2(\Omega, \mathbb{C}^3)$*

$$u(x) := (k^2 + \nabla \operatorname{div}) \iint_\Omega f_1(y) \Phi_k(x,y) \, dy + \operatorname{curl} \iint_\Omega f_2(y) \Phi_k(x,y) \, dy$$

*for $x \in \mathbb{R}^3$ defines a radiating solution of*

$$\operatorname{curl}^2 u - k^2 u = k^2 f_1 + \operatorname{curl} f_2$$

*in the weak sense and the far field pattern is given by*

$$u^\infty(\hat{x}) = k^2 \hat{x} \times \iint_\Omega f_1(y) e^{-ik\hat{x}\cdot y} \, dy \times \hat{x} + ik\,\hat{x} \times \iint_\Omega f_2(y) e^{-ik\hat{x}\cdot y} dy$$

*for $\hat{x} \in \mathbb{S}^2$.*

*Proof.* Lemma II.9 gives the first assertion. Plugging in the assymptotic behavior of $\Phi_k$ (Lemma III.1) yields the far field pattern $u^\infty$. Here we use

$$\begin{aligned}(k^2 + \nabla \operatorname{div}) &\iint_\Omega f_1(y) \Phi_k(x,y) \, d(y) \\ &= (k^2 + \operatorname{curl}^2 + \Delta) \iint_\Omega f_1(y) \Phi_k(x,y) \, d(y) \\ &= \operatorname{curl}^2 \iint_\Omega f_1(y) \Phi_k(x,y) \, d(y),\end{aligned}$$

since

$$(\Delta + k^2) \iint_\Omega f_1(y) \Phi_k(x,y) \, d(y) = 0$$

in $\mathbb{R}^3 \setminus \overline{\Omega}$. □

## 2. Factorization of the far field operator

Writing down the expression for $\mathcal{H}^*\varphi$

$$\mathcal{H}^*(\varphi_1,\varphi_2)(d) = d \times \iint_\Omega \varphi_1(y)\, e^{-ik\, d\cdot y}\, dy \times d$$
$$+ ik\, d \times \iint_\Omega \varphi_2(y)\, e^{-ik\, d\cdot y}\, dy \qquad (3.7)$$

we observe that the image $\mathcal{H}^*\varphi$ of $\varphi$ is the far field pattern $u^\infty$ of

$$u(x) = (k^2 + \nabla_x \operatorname{div}_x) \iint_\Omega \frac{1}{k^2}\varphi_1(y)\,\Phi_k(x,y)\, dy$$
$$+ \operatorname{curl} \iint_\Omega \varphi_2(y)\,\Phi_k(x,y)\, dy$$

which is the weak radiating solution of

$$\operatorname{curl}^2 u - k^2 u = \varphi_1 + \operatorname{curl}\varphi_2$$

by the previous proposition. How do we have to choose the source $\varphi$ so that the far field pattern $u^\infty$ is the same as $v^\infty = Gf$? The scattering equation for $v$ (3.6) can be written as

$$\iint_{\mathbb{R}^3} \operatorname{curl} v \cdot \operatorname{curl}\psi - k^2\, v \cdot \psi\, dx$$
$$= \iint_\Omega k^2\,[q_\mu w_1 + \mu\beta w_2]\cdot \psi\, dx \qquad (3.8)$$
$$+ \iint_\Omega \left[(q_\varepsilon + k^2 \mu\beta^2)w_2 + k^2 \mu\beta w_1\right]\cdot \operatorname{curl}\psi\, dx$$

with
$$w_1 := f_1 + v \quad \text{and} \quad w_2 = f_2 + \operatorname{curl} v.$$

Then $v^\infty$ is given by

$$v^\infty(\hat{x}) = k^2\, \hat{x} \times \iint_\Omega [q_\mu w_1 + \mu\beta w_2]\, e^{ik\hat{x}\cdot(\cdot)}\, dy \times \hat{x}$$
$$+ ik\, \hat{x} \times \iint_\Omega \left[(q_\varepsilon + k^2 \mu\beta^2)w_2 + k^2 \mu\beta w_1\right] e^{ik\hat{x}\cdot(\cdot)}\, dy.$$

Comparing this expression to the image of $\mathcal{H}^*$ (3.7) suggests to choose $\varphi = \mathcal{T}f$ as follows:

**Definition III.16 (Middle operator).** Let $v$ be the radiating solution of the transmission problem (3.6). Define the linear operator $\mathcal{T}$ by $\mathcal{T}\colon L^2(\Omega,\mathbb{C}^3)^2 \to L^2(\Omega,\mathbb{C}^3)^2$,

$$\mathcal{T}(f_1,f_2) := \begin{pmatrix} k^2 q_\mu & k^2 \mu\beta \\ k^2 \mu\beta & q_\varepsilon + k^2\mu\beta^2 \end{pmatrix} \begin{pmatrix} f_1 + v \\ f_2 + \operatorname{curl} v \end{pmatrix}.$$

**Remark III.17.** (a) The matrix–vector–multiplication is non-standard: The vector $(f_1+v, f_2+\operatorname{curl} v)^\top$ has six entries but the matrix is a $(2\times 2)$–matrix. We use the notation to abbreviate

$$\begin{pmatrix} k^2 q_\mu & k^2 \mu\beta \\ k^2 \mu\beta & q_\varepsilon + k^2\mu\beta^2 \end{pmatrix} \begin{pmatrix} f_1 + v \\ f_2 + \operatorname{curl} v \end{pmatrix}$$
$$:= \begin{pmatrix} k^2 q_\mu(f_1+v) + k^2\mu\beta(f_2+\operatorname{curl} v) \\ k^2\mu\beta(f_1+v) + (q_\varepsilon + k^2\mu\beta^2)(f_2+\operatorname{curl} v) \end{pmatrix}.$$

(b) Denote the matrix function in the definition above by $Q = Q(x)$ having in mind that the matrix represents the constrast. Then $\mathcal{T}(f) = Qw$ with $w = (w_1, w_2)^\top = (f_1+v, f_2+\operatorname{curl} v)^\top$.

(c) $\mathcal{T}$ is (bounded and) injective: $\mathcal{T}(f) = 0$ implies $w_1 \equiv 0 \equiv w_2$ in $\Omega$ and equation (3.8) yields $\operatorname{curl}^2 v - k^2 v = 0$ in $\mathbb{R}^3$. By the uniqueness assumption $v \equiv 0$ and then also $f = (f_1, f_2) \equiv 0$.

The definition of the operator $\mathcal{T}$ finishes the factorization of $\mathcal{F}$: Using the notation $\hat{\psi} := (\psi, \operatorname{curl}\psi)^\top$ for functions $\psi$ with well defined curl, $u$ satisfies

$$\iint_{\mathbb{R}^3} \operatorname{curl} u \cdot \operatorname{curl}\psi - k^2 u \cdot \psi \, \mathrm{d}x = \iint_\Omega (Qw) \cdot \hat{\psi} \, \mathrm{d}x$$
$$= \iint_{\mathbb{R}^3} \operatorname{curl} v \cdot \operatorname{curl}\psi - k^2 v \cdot \psi \, \mathrm{d}x$$

where $v$ is the unique (by assumption) radiating solution of (3.6). Here again, our notation allows to express the integral over $\Omega$ in an elegant way:

$$\iint_\Omega (Qw) \cdot \hat{\psi} \, \mathrm{d}x = \iint_\Omega \left[ k^2 q_\mu(f_1+v) + k^2\mu\beta(f_2+\operatorname{curl} v) \right] \cdot \psi \, \mathrm{d}x$$
$$+ \iint_\Omega \left[ (q_\varepsilon + k^2\mu\beta^2)(f_2+\operatorname{curl} v) + k^2\mu\beta(f_1+v) \right] \cdot \operatorname{curl}\psi \, \mathrm{d}x.$$

## 2. Factorization of the far field operator

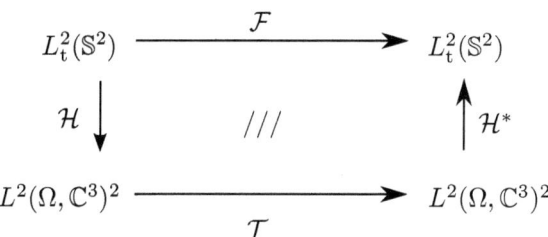

Figure III.1: Factorization of $\mathcal{F}$

Hence, $u = v$ and $G(f_1, f_2) = \mathcal{H}^*\mathcal{T}(f_1, f_2)$. Finally $\mathcal{F} = G\mathcal{H} = \mathcal{H}^*\mathcal{T}\mathcal{H}$. Figure III.1 illustrates the factorization showing the occuring function spaces and the operators between them.

### 2.1. Modified factorization

We modify the factorization of $\mathcal{F}$ for the case of non–absorbing media. For the remaining part of this section we assume that the matrix $Q$ is real–valued and (symmetric) positive definite.

If we have a look at the factorization we observe that $\mathcal{T}$ is only applied to the range of $\mathcal{H}$. The functions in $\mathcal{R}(\mathcal{H})$ admit more regularity than $L^2$ functions. Furthermore, when studying the middle operator on positivity and coercivity in the case of real valued material functions the inner product

$$\left(\mathcal{T}(f), f\right)_{L^2} = \left(Q\begin{pmatrix}f_1 + v \\ f_2 + \operatorname{curl} v\end{pmatrix}, \begin{pmatrix}f_1 \\ f_2\end{pmatrix}\right) = \left(\begin{pmatrix}f_1 + v \\ f_2 + \operatorname{curl} v\end{pmatrix}, \begin{pmatrix}f_1 \\ f_2\end{pmatrix}\right)_{L_Q^2}$$

can be interpreted as an inner product of an weighted $L^2$–space to be defined. Recall that $Q$ is symmetric positive definite. These two facts suggest to define $\mathcal{T}$ on a function space $X$ which makes use of the regularity and the weight $Q$. We define a vector version of the space $X$ introduced by Kirsch in [20].

**Definition III.18.** Given the matrix function $Q \in L^\infty(\Omega, \mathbb{R}^{2\times 2})$ such that $Q(x)$ is symmetric positve definite for almost all $x \in \Omega$. Define

(a) $L_Q^2(\Omega) := \{f \in L^2(\Omega, \mathbb{C}^3)^2 \ : \ \iint_\Omega (Qf) \cdot \overline{f} \, dx < \infty\}$

with inner product

$$(f,g)_{L_Q^2} = \iint_\Omega (Q(x)f(x)) \cdot \overline{g(x)} \, dx.$$

(b) $H_0(\text{curl}, \Omega) := \{v \in H(\text{curl}, \Omega) : \nu \times v = 0\}$,

(c) $H_{00}(\text{curl}^2, \Omega) := \{v \in H_0(\text{curl}, \Omega) : \text{curl}\, v \in H_0(\text{curl}\, v)\}$,

(d) $X := \left\{ f \in L_Q^2(\Omega) : \begin{array}{l} \iint_\Omega f \cdot (\text{curl}^2 w - k^2 w) \, dx = 0 \\ \text{f.a. } w \in H_{00}(\text{curl}^2, \Omega) \times H_{00}(\text{curl}^2, \Omega) \\ \text{with } (\text{curl}^2 w - k^2 w) \in L_{Q^{-1}}^2(\Omega) \end{array} \right\}.$

In part (c), $\text{curl}^2 w = (\text{curl}^2 w_1, \text{curl}^2 w_2)^\top$ for $w = (w_1, w_2)^\top$ with $w_j \in H_{00}(\text{curl}^2, \Omega)$, $j = 1, 2$. As in the scalar case our function space $X$ is a closed subspace of $L_Q^2(\Omega)$ and contains the range of $\mathcal{H}$. Indeed, $\mathcal{H}f \in L^2(\Omega, \mathbb{C}^3)^2$ implies $\mathcal{H}f \in L_Q^2(\Omega)$. Furthermore, by the definition of $\mathcal{H}$ we see that $\mathcal{H}f$ is an analytic solution to Maxwell's equations in $\Omega$. Hence, by Green's theorem, $\iint_\Omega (\mathcal{H}f) \cdot (\text{curl}^2 w - k^2 w) \, dx$ vanishes for all $w \in H_{00}(\text{curl}, \Omega) \times H_{00}(\text{curl}, \Omega)$.

We now deduce a modified factorization of $\mathcal{F}$ which we will use in the case of non–absorbing media: We consider the Herglotz operator as an operator $\mathcal{H} : L_t^2(\mathbb{S}^2) \to L_Q^2(\Omega)$ and denote its adjoint by $\mathcal{H}^\dagger$ with $\mathcal{H}^\dagger : L_Q^2(\Omega) \to L_t^2(\mathbb{S}^2)$ and compute:

$$(\mathcal{H}p, \varphi)_{L_Q^2(\Omega)} = (Q\mathcal{H}p, \varphi)_{L^2(\Omega,\mathbb{C}^3)^2} = (\mathcal{H}p, Q\varphi)_{L^2(\Omega,\mathbb{C}^3)^2}$$
$$= (p, \mathcal{H}^*(Q\varphi))_{L_t^2(\mathbb{S}^2)}.$$

Hence, $\mathcal{H}^\dagger$ is given by $\varphi \mapsto \mathcal{H}^*(Q\varphi)$ for $\varphi \in L_Q^2(\Omega)$ and the factorization is

$$\mathcal{F} = \mathcal{H}^\dagger Q^{-1} \mathcal{T} \mathcal{H} = \mathcal{H}^\dagger \widetilde{\mathcal{T}} \mathcal{H}$$

with $\widetilde{\mathcal{T}} : L_Q^2(\Omega) \to L_Q^2(\Omega)$, $f \mapsto (f_1 + v, f_2 + \text{curl}\, v)^\top$ where $v$ is the radiating solution of the transmission problem (3.6). The inverse matrix $Q^{-1}$ exists for almost all $x \in \Omega$ since $Q$ is symmetric positive definite. At this point, we introduce the orthogonal projector $\mathcal{P} : L_Q^2(\Omega) \hookrightarrow X$. It is a well known result from functional analysis that the null space of the

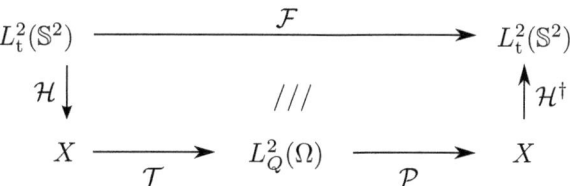

Figure III.2: Modified factorization of $\mathcal{F}$

adjoint operator $A^*$ is the orthogonal complement of the range of $A$. In our case:
$$\mathcal{N}(\mathcal{H}^\dagger) = \mathcal{R}(\mathcal{H})^\perp \supset X^\perp$$
For $\varphi \in L^2_Q(\Omega)$: $\varphi = \mathcal{P}\varphi + (I - \mathcal{P})\varphi =: \varphi_X + \varphi_{X^\perp}$ with $\varphi_X \in X$ and $\varphi_{X^\perp} \in X^\perp \subset \mathcal{N}(\mathcal{H}^\dagger)$. Then
$$\mathcal{H}^\dagger \varphi = \mathcal{H}^\dagger \varphi_X.$$
Hence, $\mathcal{H}^\dagger = \mathcal{H}^\dagger \mathcal{P}$. As mentionend above, $\widetilde{\mathcal{T}}$ is applied to functions in $\mathcal{R}(\mathcal{H}) \subset X$. We redefine $\mathcal{H} \colon L^2_t(\mathbb{S}^2) \to X$ and $\widetilde{\mathcal{T}} \colon X \to L^2_Q(\Omega)$. Then the adjoint Herglotz operator is defined on $X$ and we have the factorization
$$\mathcal{F} = \mathcal{H}^\dagger \mathcal{P} \widetilde{\mathcal{T}} \mathcal{H}.$$
Figure III.2 shows a diagram of the new situation.

## 3. Properties of the middle operator

In this section we collect important properties of the operators $\mathcal{T}$ and $\widetilde{\mathcal{T}}$, which appear in the factorization $\mathcal{F} = \mathcal{H}^* \mathcal{T} \mathcal{H}$ and $\mathcal{F} = \mathcal{H}^\dagger \mathcal{P} \widetilde{\mathcal{T}} \mathcal{H}$, respectively. Due to this we can apply abstract theorems from functional analysis to characterize the range of $\mathcal{H}^*$ and $\mathcal{H}^\dagger$ by $\mathcal{F}$. Then we are able to give a criterion for the localization of $\Omega$ depending on the range of the far field operator.

We start with the case of absorbing media; the parameter functions have non–vanishing imaginary parts. In this case $\operatorname{Im} \mathcal{T} := (\mathcal{T} - \mathcal{T}^*)/(2i)$ is coercive. In the second part we study the real valued case.

## 3.1. Absorbing media

**Theorem III.19 (Properties of $\mathcal{T}$).** *Let Assumption III.11 be satisfied. Then*

*(a)* $\operatorname{Im}\left(\mathcal{T}(f), f\right) \geq 0$ *for* $f \in L^2(\Omega, \mathbb{C}^3)^2$.

*(b) Assume that there exist constants $\gamma_\varepsilon, \gamma_\mu$ such that $\operatorname{Im} q_\varepsilon \geq \gamma_\varepsilon$ and $\operatorname{Im} q_\mu \geq \gamma_\mu$ a.e. in $\Omega$. Then $\operatorname{Im} \mathcal{T}$ is coercive on $\mathcal{R}(\mathcal{H})$; that is, there exists a constant $\gamma > 0$ such that*

$$\operatorname{Im}\left(\mathcal{T}(f), f\right) \geq \gamma \|f\|_{L^2}^2$$

*for all $f \in \mathcal{R}(\mathcal{H})$.*

*Proof.* The properties are shown analogously to the non–magnetic achiral case in [24]. For $f \in L^2(\Omega, \mathbb{C}^3)^2$, and a solution $v$ of (3.6) $\mathcal{T}(f) = Qw$ with

$$Q = \begin{pmatrix} k^2 q_\mu & k^2 \mu \beta \\ k^2 \mu \beta & q_\varepsilon + k^2 \mu \beta^2 \end{pmatrix}$$

and $w = (w_1, w_2)^\top$ where $w_1 = f_1 + v$ and $w_2 = f_2 + \operatorname{curl} v$. Remember that (3.6) is equivalent to (3.8); that is,

$$\iint_{\mathbb{R}^3} \operatorname{curl} v \cdot \operatorname{curl} \psi - k^2 v \cdot \psi \, dx = \iint_\Omega (Qw) \cdot \hat{\psi} \, dx.$$

We compute

$$\left(\mathcal{T}(f), f\right)_{L^2(\Omega, \mathbb{C}^3)^2} = \left(Qw, \overline{(w - \hat{v})}\right)_{L^2(\Omega, \mathbb{C}^3)^2}$$

$$= \iint_\Omega q_\varepsilon |w_2|^2 + k^2 \left[\mu \beta^2 |w_2|^2 + 2\mu \beta \operatorname{Re}(w_1 \cdot \overline{w}_2) + q_\mu |w_1|^2\right] dx$$

$$- \iint_\Omega (Qw) \cdot \hat{\overline{v}} \, dx.$$

We first look at the last integral. We choose $R > 0$ such that $\overline{\Omega}$ is contained in the ball with radius $R$. We set $\psi = \phi \overline{v}$ in (3.8) where $\phi \in \mathcal{C}^\infty(\mathbb{R}^3)$ is a cutoff function with $\phi \equiv 1$ for $|x| \leq R$ and $\phi \equiv 0$ for $|x| \geq 2R$ (compare Figure III.3). This yields:

## 3. Properties of the middle operator

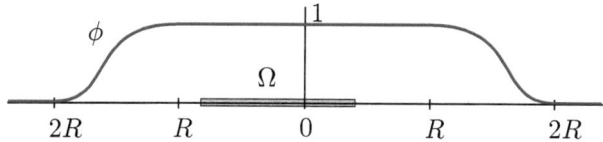

Figure III.3: Cutoff function $\phi$

$$\iint_\Omega (Qw) \cdot \hat{\overline{v}} \, dx$$
$$= \iint_{|x|<R} |\operatorname{curl} v|^2 - k^2|v|^2 \, dx + \iint_{R<|x|<2R} \operatorname{curl} v \cdot \overline{\operatorname{curl}(\overline{v}\phi)} - k^2|v|^2 \phi \, dx$$
$$= \iint_{|x|<R} |\operatorname{curl} v|^2 - k^2|v|^2 \, dx + \int_{|x|=R} (\hat{x} \times \operatorname{curl} v) \cdot \overline{v} \, ds$$

by Green's Theorem. Taking the imaginary part yields

$$\operatorname{Im}\left(\mathcal{T}(f), f\right)_{L^2(\Omega,\mathbb{C}^3)^2}$$
$$= \operatorname{Im} \iint_\Omega q_\varepsilon |w_2|^2 + k^2\left[\mu\beta^2|w_2|^2 + 2\mu\beta \operatorname{Re}(w_1 \cdot \overline{w_2}) + q_\mu |w_1|^2\right] dx$$
$$- \operatorname{Im} \int_{|x|=R} (\hat{x} \times \operatorname{curl} v(x)) \cdot \overline{v(x)} \, ds(x).$$

From the radiation condition

$$\lim_{|x|\to\infty} |x| \big(\operatorname{curl} v(x) \times \hat{x} - ik\, v(x)\big) = 0$$

and the far field expansion we conclude the limit

$$\lim_{R\to\infty} \int_{|x|=R} (\hat{x} \times \operatorname{curl} v(x)) \cdot \overline{v(x)} \, ds(x) = -\frac{ik}{(4\pi)^2} \int_{\mathbb{S}^2} |v^\infty|^2 \, ds.$$

Finally, using the binomial $|a+b|^2 = |a|^2 + 2\operatorname{Re}(a \cdot \overline{b}) + |b|^2$ for vectors $a,b \in \mathbb{C}^3$ and the fact that $\operatorname{Im}\mu = \operatorname{Im} q_\mu$ we derive

$$\operatorname{Im}\left(\mathcal{T}(f), f\right) = \iint_\Omega \operatorname{Im} q_\varepsilon |w_2|^2 + k^2 \operatorname{Im} q_\mu \left|\beta w_2 + w_1\right|^2 dx + \frac{k}{(4\pi)^2} \int_{\mathbb{S}^2} |v^\infty|^2 \, ds.$$

(a) From the above formula for $\operatorname{Im}(\mathcal{T}f,f)$ we see that $\operatorname{Im}(\mathcal{T}f,f) \geq 0$ if the imaginary parts of the contrasts are non–negative. But that is the case, as $\operatorname{Im} q_\mu = \operatorname{Im} \mu \geq 0$ and $\operatorname{Im} q_\varepsilon = \operatorname{Im} \varepsilon/|\varepsilon|^2 \geq 0$ by our Assumption III.11.

(b) If there exists no such $\gamma$ one can find a sequence $\{f^j\} \subset L^2(\Omega, \mathbb{C}^3)^2$ s.t. $\|f^j\| = 1$ and $\operatorname{Im}(\mathcal{T}f^j, f^j) \to 0$. For each $f^j$ there is a corresponding $v_j$ and also $w^j = (w_1^{(j)}, w_2^{(j)})$. The above formula shows that with $\operatorname{Im}(\mathcal{T}f^j, f^j) \to 0$ also $w_2^j \to 0$ and $\beta w_2^j + w_1^j \to 0$, so also $w_1^j \to 0$. By the unique solvability $v_j$ converges to 0 in $H(\operatorname{curl}, \Omega)$. And therefore $f_1^j = w_1^j - v_j$ and $f_2^j = w_2^j - \operatorname{curl} v_j$ converge to 0 in $L^2(\Omega, \mathbb{C}^3)$ which is a contradiction to $\|f^j\| = 1$. □

**Corollary III.20.** *Let Assumption III.11 be satisfied. Then the operator* $\operatorname{Im}\mathcal{F} := (\mathcal{F} - \mathcal{F}^*)/(2i)$ *is self–adjoint, compact and positive. Furthermore, in the case of part (b); that is,* $\operatorname{Im}\mathcal{T}$ *coercive (in the sense of the previous theorem),* $\operatorname{Im}\mathcal{F}$ *is injective. In that case there exists a complete orthonormal eigensystem* $\{\lambda_j, \psi_j\}_{j\in\mathbb{N}}$ *with positive eigenvalues* $\lambda_j$ *and*

$$(\operatorname{Im}\mathcal{F})p = \sum_{j\in\mathbb{N}} \lambda_j (p, \psi_j)\psi_j.$$

*Proof.* By definition $\operatorname{Im}\mathcal{F}$ is self–adjoint. $\mathcal{F}$ is compact, thus also $\operatorname{Im}\mathcal{F}$. By the factorization for any $p \in L_t^2(\mathbb{S}^2)$

$$\big((\operatorname{Im}\mathcal{F})p, p\big) = \big(\mathcal{H}^*(\operatorname{Im}\mathcal{T})\mathcal{H}p, p\big) = \big((\operatorname{Im}\mathcal{T})\mathcal{H}p, \mathcal{H}p\big) = \operatorname{Im}(\mathcal{T}\mathcal{H}p, \mathcal{H}p) \geq 0$$

by part (a), whence the positivity. Given $p \in L_t^2(\mathbb{S}^2)$ such that $\operatorname{Im}\mathcal{F}p = 0$ we have by part (b)

$$0 = \big((\operatorname{Im}\mathcal{F})p, p\big) = \big((\operatorname{Im}\mathcal{T})\mathcal{H}p, \mathcal{H}p\big) \geq \gamma\|\mathcal{H}p\|^2.$$

Hence $\mathcal{H}p = 0$. Since $\mathcal{H}$ is injective, this implies that $p = 0$. The spectral theorem for compact self–adjoint operators yields the existence of the eigensystem and the series representation. □

## 3.2. Non–absorbing media

In this subsection we consider the case of real valued parameter functions $\varepsilon, \mu$ and $\beta$ characterizing the chiral material. The main assumption of the

## 3. Properties of the middle operator

previous theorem was an estimation for the contrasts $q_\varepsilon$, $q_\mu$ of the form $\operatorname{Im} q \geq \gamma$. We were able to prove coercivity of $\operatorname{Im} \mathcal{T}$. Furthermore, by the fact that $\operatorname{Im} q_\mu = \operatorname{Im} \mu$ we could exploit the concept of completing the square. These two items prevent us from using the same approach in the real valued case.

We choose an alternative approach and use the modified factorization

$$\mathcal{F} = \mathcal{H}^\dagger \mathcal{P} \widetilde{\mathcal{T}} \mathcal{H}$$

with operators $\mathcal{H}: L_t^2(\mathbb{S}^2) \to X$, $\widetilde{\mathcal{T}}: X \to L_Q^2(\Omega)$ and $\mathcal{P}: L_Q^2(\Omega) \hookrightarrow X$. For the definition of the function space $X$ we refer to III.18.

We will show that $\widetilde{\mathcal{T}}$ is the sum of a coercive operator $\mathcal{T}_0$ and a compact one: $\mathcal{T}_0$ is coercive on $X$ and $\widetilde{\mathcal{T}} - \mathcal{T}_0$ is compact from $X$ into $L_Q^2(\Omega)$. Then also $\mathcal{P}\mathcal{T}_0$ is coercive on $X$ and $\mathcal{P}(\widetilde{\mathcal{T}} - \mathcal{T}_0)$ is compact from $X$ into $X$.

We can treat two cases: Firstly $\varepsilon, \mu > 1$ then the contrasts $q_\varepsilon$ and $q_\mu$ are positive and secondly $0 < \varepsilon, \mu < 1$. Then the contrasts are negative.

### Positive contrasts

Recall the symmetric matrix function $Q = Q(x)$,

$$Q = \begin{pmatrix} k^2 q_\mu & k^2 \mu \beta \\ k^2 \mu \beta & q_\varepsilon + k^2 \mu \beta^2 \end{pmatrix}.$$

By our assumptions on the parameter functions, $Q$ has compact support in $\overline{\Omega}$. We have to make some smoothness assumptions on $Q$ and follow [20]. Alternatively, Lechleiter treats the achiral Maxwell's equations in [27] with constrasts in $\mu$ and $\varepsilon$ and uses integrability assumptions on the contrasts.

Notation: For sufficiently smooth $Q = (q_{jl})_{j,l=1,2}$ define the matrix

$$\nabla Q := (\nabla q_{jl})_{j,l=1,2}.$$

The following product rule holds for $\psi = (\psi_1, \psi_2) \in H^1(\Omega, \mathbb{C}) \times H^1(\Omega, \mathbb{C})$:

$$\nabla(Q\psi) = (\nabla Q)\psi + Q\nabla\psi$$

where $\nabla \psi := (\nabla \psi_1, \nabla \psi_2)^\top$. Indeed:

$$\nabla(Q\psi) = \nabla \begin{pmatrix} q_{11}\psi_1 + q_{12}\psi_2 \\ q_{21}\psi_1 + q_{22}\psi_2 \end{pmatrix} = \begin{pmatrix} \nabla q_{11}\psi_1 + q_{11}\nabla\psi_1 + \nabla q_{12}\psi_2 + q_{12}\nabla\psi_2 \\ \nabla q_{21}\psi_1 + q_{21}\nabla\psi_1 + \nabla q_{22}\psi_2 + q_{22}\nabla\psi_2 \end{pmatrix}$$

$$= \begin{pmatrix} \nabla q_{11} & \nabla q_{12} \\ \nabla q_{21} & \nabla q_{22} \end{pmatrix} \begin{pmatrix} \psi_1 \\ \psi_2 \end{pmatrix} + \begin{pmatrix} q_{11} & q_{12} \\ q_{21} & q_{22} \end{pmatrix} \begin{pmatrix} \nabla\psi_1 \\ \nabla\psi_2 \end{pmatrix}.$$

Here again, the matrix–vector–products in the second line are non–standard.

**Assumption III.21.** *Let $\Omega \subset \mathbb{R}^3$ be a bounded Lipschitz domain such that the complement $\mathbb{R}^3 \setminus \overline{\Omega}$ is connected. Let $k > 0$ be the wave number and $\varepsilon, \mu \in L^\infty(\Omega, \mathbb{R})$ and $\beta \in L^\infty(\Omega, \mathbb{R})$ real valued. We extend $\varepsilon$ and $\mu$ by one and $\beta$ by zero outside of $\Omega$. We assume:*

*(a) $Q$ is (symmetric) positive definite.*

*(b) $Q \in C(\mathbb{R}^3, \mathbb{R}^{2\times 2})$, $Q|_\Omega \in C^1(\Omega, \mathbb{R}^{2\times 2})$, $Q^{-1/2}(\nabla Q) \in L^\alpha(\Omega, \mathbb{R}^{6\times 2})$ for some $\alpha > 3$.*

*(c) For all $(g, h) \in L^2(\Omega, \mathbb{C}^3) \times L^2(\Omega, \mathbb{C}^3)$ there exists a unique radiating solution of the transmission problem (2.20) for $\kappa = k$.*

*(d) There exists a constant $c > 0$ such that $\frac{1}{\varepsilon} - 2k^2\mu\beta^2 \geq c$ a.e. in $\Omega$.*

$Q^{-1/2} =: (\tilde{q}_{jl})_{j,l=1,2}$ is well defined and

$$Q^{-1/2}(\nabla Q) = \begin{pmatrix} \tilde{q}_{11} & \tilde{q}_{12} \\ \tilde{q}_{21} & \tilde{q}_{22} \end{pmatrix} \begin{pmatrix} \nabla q_{11} & \nabla q_{12} \\ \nabla q_{21} & \nabla q_{22} \end{pmatrix}$$

$$:= \begin{pmatrix} \tilde{q}_{11}\nabla q_{11} + \tilde{q}_{12}\nabla q_{21} & \tilde{q}_{11}\nabla q_{12} + \tilde{q}_{12}\nabla q_{22} \\ \tilde{q}_{21}\nabla q_{11} + \tilde{q}_{22}\nabla q_{21} & \tilde{q}_{21}\nabla q_{12} + \tilde{q}_{22}\nabla q_{22} \end{pmatrix}.$$

Under which conditions is $Q$ positive definite? We compute

$$\xi^\top Q \xi = k^2 q_\mu |\xi_1|^2 + 2k^2 \mu\beta \mathrm{Re}\,(\xi_1\overline{\xi_2}) + k^2\mu\beta^2|\xi_2|^2 + q_\varepsilon|\xi_2|^2$$

$$= k^2 q_\mu \left|\xi_1 + \left(1 - \tfrac{\mu\beta}{q_\mu}\right)\xi_2\right|^2 + q_\varepsilon\left(1 - k^2 \tfrac{\mu\beta^2}{q_\varepsilon q_\mu}\right)|\xi_2|^2.$$

$Q$ is positive definite if $q_\mu > 0$, $q_\varepsilon > 0$ and $1 - k^2\mu\beta^2/(q_\varepsilon q_\mu) > 0$. The last inequality can be interpreted as a restriction for $k^2\beta^2$: this quantity should be sufficiently small. The same applies to part (d). Furthermore,

## 3. Properties of the middle operator

the fractions $\frac{\mu\beta}{q_\mu}$ and $\frac{\mu\beta^2}{q_\varepsilon q_\mu}$ may not be singular on the boundary: Therefore $\beta$ and $\beta^2$ must decay at least as fast as $q_\mu$ and $q_\varepsilon q_\mu$, respectively.

Before we study the middle operator we give a vector version of the auxiliary lemma 5.3 in [20]:

**Lemma III.22.** *Assume that $Q$ satisfies part (a) and (b) of the above Assumption III.21. Let $\psi = (\psi_1, \psi_2)^\top \in H^1(\Omega) \times H^1(\Omega)$. Then*

*(a) $Q^{-1/2}(\nabla Q)\psi \in L^2(\Omega, \mathbb{C}^3) \times L^2(\Omega, \mathbb{C}^3)$ and*

*(b) $\nabla(Q\psi) \in H_{00}(\mathrm{curl}^2, \Omega) \times H_{00}(\mathrm{curl}^2, \Omega)$.*

The proof of the lemma is just the vector version of the scalar case. The vector $Q^{-1/2}(\nabla Q)\psi$ consists of terms which satisfy the assumptions of Lemma 5.3 in [20]. For part (b) we apply Lemma 5.3 to both components of $\nabla(Q\psi)$.

For vector functions $\psi$ with well–defined curl define the vector

$$\hat{\psi} := \begin{pmatrix} \psi \\ \mathrm{curl}\,\psi \end{pmatrix}.$$

Recall the definition of $\widetilde{\mathcal{T}} \colon X \to L^2_Q(\Omega)$:

$$\widetilde{\mathcal{T}} f = \begin{pmatrix} f_1 + v \\ f_2 + \mathrm{curl}\, v \end{pmatrix} = f + \hat{v}$$

where $v \in H_{\mathrm{loc}}(\mathrm{curl}, \mathbb{R}^3)$ is the radiating solution of (3.6) namely

$$\iint_{\mathbb{R}^3} \left[ \left(\tfrac{1}{\varepsilon} - k^2 \mu \beta^2\right) \mathrm{curl}\, v - k^2 \mu \beta v \right] \cdot \mathrm{curl}\,\psi - k^2 \left[\mu \beta\, \mathrm{curl}\, v + \mu v\right] \cdot \psi\, \mathrm{d}x$$

$$= \iint_\Omega (Qf) \cdot \hat{\psi}\, \mathrm{d}x.$$

Let $v_0$ be the solution of a slightly different equation

$$\iint_{\mathbb{R}^3} \left[ \left(\tfrac{1}{\varepsilon} - k^2 \mu \beta^2\right) \mathrm{curl}\, v_0 - k^2 \mu \beta v_0 \right] \cdot \mathrm{curl}\,\psi - k^2 \left[\mu \beta\, \mathrm{curl}\, v_0 - \mu v_0\right] \cdot \psi\, \mathrm{d}x$$

$$= \iint_\Omega (Qf) \cdot \hat{\psi}\, \mathrm{d}x \tag{3.9}$$

and define $\mathcal{T}_0 \colon X \to L^2_Q(\Omega)$ by

$$\mathcal{T}_0 f = f + \hat{v}_0. \qquad (3.10)$$

Then $\widetilde{\mathcal{T}} f = \mathcal{T}_0 f + (\widetilde{\mathcal{T}} - \mathcal{T}_0) f$ and $(\widetilde{\mathcal{T}} - \mathcal{T}_0) f = \hat{v} - \hat{v}_0$.

As a first step, we give an existence and uniqueness result for $v_0$ and show that the mapping $f \mapsto v_0$ is compact in $L^2(\Omega, \mathbb{C}^3)$ – due to our smoothness assumptions on $Q$.

**Proposition III.23.** *Let Assumption III.21 be satisfied.*

*(a) The variational equation (3.9) for $v_0$ is uniquely solvable.*

*(b) The mapping $f \mapsto v_0|_\Omega$ is compact from $X$ into $L^2_Q(\Omega)$.*

*Proof.* (a) This proof is very similar to the proof of part (c) of the corresponding Theorem II.17. As in the proof of Theorem II.10, the variational problem for $v_0$ is equivalent to the IDE by Lemma II.9

$$\begin{aligned}v_0(x) = (-k^2 + \nabla \operatorname{div}) &\iint_\Omega \left[ q_\mu v_0 - \mu\beta \operatorname{curl} v_0 - g \right] \Phi_{ik}(x, \cdot) \, dy \\ + \operatorname{curl} &\iint_\Omega \left[ (q_\varepsilon + k^2\mu\beta^2) \operatorname{curl} v_0 + k^2\mu\beta v_0 + h \right] \Phi_{ik}(x, \cdot) \, dy\end{aligned} \qquad (3.11)$$

where

$$k^2 g = (Qf)_1 = k^2 q_\mu f_1 + k^2 \mu\beta f_2$$

and

$$h = (Qf)_2 = (q_\varepsilon + k^2\mu\beta^2) f_2 + k^2 \mu\beta f_1.$$

From this we see that $v_0(x)$ decays exponentially as $|x| \to \infty$. Thus we define a sesqui–linear form on $H(\operatorname{curl}, \mathbb{R}^3) \times H(\operatorname{curl}, \mathbb{R}^3)$ and a conjugate–linear form on $H(\operatorname{curl}, \mathbb{R}^3)$.

$$\begin{aligned}a(w, \psi) &:= \iint_{\mathbb{R}^3} \left( \tfrac{1}{\varepsilon} - k^2 \mu\beta^2 \right) \operatorname{curl} w \cdot \overline{\operatorname{curl} \psi} \, dx \\ &\quad + k^2 \iint_{\mathbb{R}^3} \mu w \cdot \overline{\psi} - \mu\beta w \cdot \overline{\operatorname{curl} \psi} - \mu\beta \operatorname{curl} w \cdot \overline{\psi} \, dx, \\ b(\psi) &:= \iint_\Omega k^2 g \cdot \overline{\psi} + h \cdot \overline{\operatorname{curl} \psi} \, dx.\end{aligned}$$

## 3. Properties of the middle operator

$a$ and $b$ are obviously bounded. We show coercivity of $a$.

$$a(w,w) = \iint_{\mathbb{R}^3} \left(\tfrac{1}{\varepsilon} - 2k^2\mu\beta^2\right) \operatorname{curl} w \cdot \operatorname{curl} \overline{w} + k^2\mu\beta^2 \operatorname{curl} w \cdot \operatorname{curl} \overline{w} \, dx$$

$$+ k^2 \iint_{\mathbb{R}^3} \mu w \cdot \overline{w} - \mu \beta w \cdot \operatorname{curl} \overline{w} - \mu \beta \operatorname{curl} w \cdot \overline{w} \, dx$$

$$= \iint_{\mathbb{R}^3} \left(\tfrac{1}{\varepsilon} - 2k^2\mu\beta^2\right) |\operatorname{curl} w|^2 + k^2 \mu |\beta \operatorname{curl} w - w|^2 \, dx$$

$$\geq \min\{c,1\} \, \|w\|_\beta^2$$

where $\|w\|_\beta^2 := \|\operatorname{curl} w\|_{L^2(\mathbb{R}^3,\mathbb{C}^3)}^2 + \|\beta \operatorname{curl} w + w\|_{L^2(\mathbb{R}^3,\mathbb{C}^3)}^2$ is an equivalent norm to $\|\cdot\|_{H(\operatorname{curl},\mathbb{R}^3)}$. The application of the Lax–Milgram lemma finishes the proof.

(b) We prove this assertion in three steps: Let the sequence $(f^n)_n \subset X$ converge weakly to zero in $L^2_Q(\Omega)$ and let $(v_0^n)_n$ the corresponding sequence of solutions. The solution operator is bounded. Hence, for any ball $B \subset \mathbb{R}^3$ with $\overline{\Omega} \subset B$ the sequence $(v_0^n)_n$ converges weakly in $H(\operatorname{curl}, B)$ and from (3.11) we conlude that $v_0^n$ is smooth in the exterior of $\Omega$ and converges uniformly to zero on the boundary $\partial B$.

We need the space of functions with well defined divergence: Let $D$ a Lipschitz domain. A function $v \in L^2(D, \mathbb{C}^3)$ admits a weak divergence if there exists a function $w \in L^2(D)$ such that

$$\iint_D v \cdot \nabla \psi + w \psi \, dx = 0$$

for all $\psi \in C_0^\infty(D)$. We write $\operatorname{div} v := w$ and denote the space of such functions by $H(\operatorname{div}, D)$.

(i) Show that $\mu v_0^n \in H(\operatorname{div}, B)$ for any ball $B \supset \overline{\Omega}$.

(ii) Determine $q^n \in H^1(B)$ such that $\frac{\partial q^n}{\partial \nu} = \nu \cdot v_0^n$ on $\partial B$ and conclude that $\|q^n\|_{H^1(B)} \to 0$.

(iii) Conclude that $\tilde{v}^n = v_0^n - \nabla q^n$ converges strongly.

(i) Let the ball $B \supset \overline{\Omega}$. For any test function $\psi = \nabla \varphi$ with $\varphi \in H_0^1(B)$ equation (3.9) yields:

$$k^2 \iint_B \mu v_0^n \cdot \nabla \varphi \, dx = k^2 \iint_\Omega \mu \beta \operatorname{curl} v_0^n \cdot \nabla \varphi \, dx + \iint_\Omega \widehat{(Qf^n) \cdot \nabla \varphi} \, dx$$

$\mu\beta \in C^1(\Omega)$ by our assumptions on $Q$. Hence $\mu\beta\operatorname{curl} v_0^n \in H(\operatorname{div},\Omega)$ since $\operatorname{div}\operatorname{curl} v_0^n = 0$. Using the divergence theorem we rewrite the first integral on the right hand side:

$$\iint_\Omega \mu\beta \operatorname{curl} v_0^n \cdot \nabla\varphi \, dx$$

$$= -\iint_\Omega \nabla(\mu\beta) \cdot \operatorname{curl} v_0^n \, \varphi \, dx + \int_{\partial\Omega} \mu\beta \, (\nu \cdot \operatorname{curl} v_0^n) \, \varphi \, ds$$

$$= -\iint_\Omega \nabla(\mu\beta) \cdot \operatorname{curl} v_0^n \, \varphi \, dx$$

For functions in $H(\operatorname{div},\Omega)$ the normal trace is well defined in the distributional sense. The boundary integral vanishes because $\mu\beta = 0$ on the boundary $\partial\Omega$. The second integral on the right hand side is treated as in the proof of Theorem 5.2 in Kirsch [20]:

$$\iint_\Omega (Qf^n) \cdot \widehat{\nabla\varphi} \, dx = -\iint_\Omega ((\nabla Q) f^n) \cdot \begin{pmatrix}\varphi\\0\end{pmatrix} dx + \underbrace{\iint_\Omega \nabla\left(Q\begin{pmatrix}\varphi\\0\end{pmatrix}\right) \cdot f^n \, dx}_{(\star)}$$

$$= -\iint_\Omega ((\nabla Q) f^n) \cdot \begin{pmatrix}\varphi\\0\end{pmatrix} dx.$$

With the aid of Lemma III.22 we conclude that the intgral $(\star)$ vanishes as in the proof of Theorem 5.2 in [20]. Plugging these results into the variational equation yields

$$k^2 \iint_B \mu v_0^n \cdot \nabla\varphi \, dx = -\iint_\Omega \left[k^2 \nabla(\mu\beta) \cdot \operatorname{curl} v_0^n + ((\nabla Q) f_n)_1\right] \varphi \, dx$$

for all $\varphi \in H_0^1(B)$. Hence, $\mu v_0^n \in H(\operatorname{div},B)$.

(ii) Now we determine $q^n \in H^1(B)$ such that

$$\iint_B \mu \nabla q^n \cdot \nabla\overline\varphi + q^n \overline\varphi \, dx = \int_{\partial B} \nu \cdot v_0^n \, \overline\varphi \, ds$$

for all $\varphi \in H^1(B)$. This Neumann problem is uniquely solvable and

$$\|q^n\|_{H^1(B)}^2 \le \iint_B \mu |\nabla q^n|^2 + |q^n|^2 \, dx = \int_{\partial B} \nu \cdot v_0^n \, \overline{q^n} \, ds$$

$$\le c \|v_0^n\|_{C(\partial B)} \|q^n\|_{H^1(B)},$$

## 3. Properties of the middle operator

since $\mu > 1$. Hence, $\|q^n\|_{H^1(B)} \leq c \|v_0^n\|_{C(\partial B)} \to 0$ $(n \to \infty)$.

(iii) Define $\tilde{v}^n := v_0^n - \nabla q^n$. Then we have: $\tilde{v}^n$ and $\operatorname{curl} \tilde{v}^n$ converge weakly to zero in $L^2(B, \mathbb{C}^3)$ and $\mu \tilde{v}^n \in H(\operatorname{div}, B)$ and $\nu \cdot \tilde{v}^n = 0$. By Weber [39] $\tilde{v}^n$ converges stongly to zero and therefore $v_0^n$ also converges strongly since $\|\nabla q^n\|_{L^2} \to 0$. □

Part (b) can also be proven for the orginial scattering problem (3.6). We formulate this result as corollary.

**Corollary III.24.** *The mapping $f \mapsto v|_\Omega$ where $v$ is the radiating solution of (3.6) is compact from $X$ into $L^2_Q(\Omega)$.*

The next theorem proves coercivity of $\mathcal{T}_0$ and that the operator $\widetilde{\mathcal{T}} - \mathcal{T}_0$ is compact. The difference $u := v - v_0$ solves

$$\iint_{\mathbb{R}^3} \left[\left(\tfrac{1}{\varepsilon} - k^2\mu\beta^2\right) \operatorname{curl} u - k^2\mu\beta u\right] \cdot \operatorname{curl}\psi - k^2 \left[\mu\beta \operatorname{curl} u + \mu u\right] \cdot \psi \, dx$$

$$= 2k^2 \iint_{\mathbb{R}^3} \mu v_0 \cdot \psi \, dx. \qquad (3.12)$$

**Theorem III.25.** *Let Assumption III.21 be satisfied and assume that $1 - k^2\varepsilon\mu\beta^2 \neq 0$ a.e.*

*(a) The operator $\mathcal{T}_0$ is self–adjoint and coercive on $X$; that is, there exists a constant $c > 0$ such that*

$$(\mathcal{T}_0 f, f)_{L^2_Q} \geq c\|f\|^2_{L^2_Q} \quad \text{for all } f \in X.$$

*(b) The operator $\widetilde{\mathcal{T}} - \mathcal{T}_0$ is compact from $X$ into $L^2_Q(\Omega)$.*

*Proof.* (a) (i) $\mathcal{T}_0$ is self–adjoint. Let $g, h \in X$ and let $v_g, v_h$ the corresponding solution of the variational problem (3.9) for $g$ and $h$, respectively. Then

$$(\mathcal{T}_0 g, h)_{L^2_Q} = (g, h)_{L^2_Q} + (Q\hat{v}_g, h)_{L^2}.$$

Recall that the matrix $Q$ is self–adjoint. It is left to consider the second term which gives the right–hand side of the variational equation (3.9) Analogously,

$$(g, \mathcal{T}_0 h)_{L^2_Q} = (g, h)_{L^2_Q} + (g, Q\hat{v}_h)_{L^2}.$$

But as equation (3.9) is symmetric in $\psi$ and $v_0$ both expressions are equal: $(Q\hat{v}_g, h)_{L^2} = (g, Q\hat{v}_h)_{L^2}$. Hence $\mathcal{T}_0$ is self–adjoint.

(ii) $\mathcal{T}_0$ is coercive: Let $f \in X$ and $v_0$ the solution of (3.9). (Recall that $v_0$ decays exponentially.)

$(\mathcal{T}_0 f, f)_{L^2_Q}$

$= (f,f)_{L^2_Q} + \iint_\Omega (Q\overline{f}) \cdot \hat{v}_0 \, dx$

$= (f,f)_{L^2_Q} + \iint_{\mathbb{R}^3} \left(\frac{1}{\varepsilon} - 2k^2\mu\beta^2\right) |\operatorname{curl} v_0|^2 + k^2\mu |\beta \operatorname{curl} v_0 + v_0|^2 \, dx$

$\geq \|f\|^2_{L^2_Q}.$

(b) (i) The mapping $f \mapsto u|_\Omega$ with $u = v - v_0$ is compact from $X$ into $L^2_Q(\Omega)$ by Proposition III.23 and Corollary III.24.

(ii) Show that the mapping $f \mapsto \operatorname{curl} u|_\Omega$ is compact: Let $B_1, B_2$ two balls with $B_1 \supset B_2 \supset \overline{\Omega}$. Let $(f^n)_n$ be a sequence in $X$ converging weakly to zero. Then the corresponding solution $(u^n)_n$ converges to zero in $L^2_Q(\Omega)$. For $n \in \mathbb{N}$ choose $\psi^n := \phi \overline{u^n}$ in equation (3.12) where $\phi \in \mathcal{C}^\infty_0(\mathbb{R}^3)$ is a cutoff function with $\phi \equiv 1$ in $\overline{B_2}$ and $\phi \equiv 0$ in $\mathbb{R}^3 \setminus B_1$. Then (3.12) reads

$\iint_{B_2} \left(\frac{1}{\varepsilon} - k^2\mu\beta^2\right) |\operatorname{curl} u^n|^2 - k^2\mu\beta(u^n \cdot \operatorname{curl} \overline{u^n} + \operatorname{curl} u^n \cdot \overline{u^n}) - k^2\mu |u^n|^2 \, dx$

$= -\iint_{B_1 \setminus B_2} \operatorname{curl} u^n \cdot \operatorname{curl}(\phi \overline{u^n}) - k^2 u^n \cdot (\phi \overline{u^n}) \, dx$

$\quad + 2k^2 \iint_{B_1} \mu v_0^n \cdot \phi \overline{u^n} \, dx$

$\iff$

$\iint_{B_2} \left(\frac{1}{\varepsilon} - k^2\mu\beta^2\right) \left|\operatorname{curl} u^n - \frac{k^2\mu\varepsilon\beta}{1 - k^2\mu\varepsilon\beta^2} u^n\right|^2 - \frac{k^2\mu}{1 - k^2\mu\varepsilon\beta^2} |u^n|^2 \, dx$

$= -\iint_{B_1 \setminus B_2} \operatorname{curl} u^n \cdot \operatorname{curl}(\phi \overline{u^n}) - k^2 u^n \cdot (\phi \overline{u^n}) \, dx$

$\quad + 2k^2 \iint_{B_1} \mu v_0^n \cdot \phi \overline{u^n} \, dx.$

## 3. Properties of the middle operator

We proceed with an application of Green's formula to the first term on the right–hand side:

$$-\iint_{B_1\setminus B_2} \operatorname{curl} u^n \cdot \operatorname{curl}(\phi\,\overline{u^n}) - k^2 u^n \cdot (\phi\,\overline{u^n})\,\mathrm{d}x$$

$$= -\iint_{B_1\setminus B_2} (\operatorname{curl}^2 - k^2) u^n \cdot (\phi\,\overline{u^n})\,\mathrm{d}x + \int_{\partial B_2} (\nu \times \overline{u^n}) \cdot \operatorname{curl} u^n\,\mathrm{d}s$$

$$= -\iint_{B_1\setminus B_2} 2k^2 v_0^n \cdot (\phi\,\overline{u^n})\,\mathrm{d}x + \int_{\partial B_2} (\nu \times \overline{u^n}) \cdot \operatorname{curl} u^n\,\mathrm{d}s.$$

Thus

$$\iint_{B_2} \left(\tfrac{1}{\varepsilon} - k^2\mu\beta^2\right)\left|\operatorname{curl} u^n - \frac{k^2\mu\varepsilon\beta}{1-k^2\mu\varepsilon\beta^2}u^n\right|^2 - \frac{k^2\mu}{1-k^2\mu\varepsilon\beta^2}|u^n|^2\,\mathrm{d}x$$

$$= 2k^2 \iint_{B_2} \mu v_0^n \cdot (\phi\,\overline{u^n})\,\mathrm{d}x + \int_{\partial B_2} (\nu \times \overline{u^n}) \cdot \operatorname{curl} u^n\,\mathrm{d}s.$$

Then we estimate the right–hand side: There exists a constant $C > 0$ such that

$$|\text{r.h.s.}| \leq C\|v_0^n\|\|u^n\| + \left|\int_{\partial B_2} (\nu \times \overline{u^n}) \cdot \operatorname{curl} u^n\,\mathrm{d}s\right| \stackrel{n\to\infty}{\longrightarrow} 0.$$

This yields

$$\iint_{B_2} \left(\tfrac{1}{\varepsilon} - k^2\mu\beta^2\right)\left|\operatorname{curl} u^n - \frac{k^2\mu\varepsilon\beta}{1-k^2\mu\varepsilon\beta^2}u^n\right|^2 - \frac{k^2\mu}{1-k^2\mu\varepsilon\beta^2}|u^n|^2\,\mathrm{d}x \to 0,$$

whence $\|\operatorname{curl} u^n\| \to 0$. $\square$

### Negative contrasts

In the case of negative constrasts the matrix $-Q$ is symmetric positive definite. We have to work in the space $L^2_{-Q}(\Omega)$ with inner product

$$(\psi, \phi)_{L^2_{-Q}} = -(Q\psi, \phi)_{L^2}.$$

The middle operator $\widetilde{\mathcal{T}}$ is then given by

$$\widetilde{\mathcal{T}}f = -(f + \hat{v}).$$

Here again for $f \in X$, $v \in H_{\mathrm{loc}}(\mathrm{curl}, \mathbb{R}^3)$ is the radiating solution of (3.6) namely

$$\iint_{\mathbb{R}^3} \left[\left(\tfrac{1}{\varepsilon} - k^2\mu\beta^2\right) \mathrm{curl}\, v - k^2\mu\beta v\right] \cdot \mathrm{curl}\, \psi - k^2\left[\mu\beta\,\mathrm{curl}\, v + \mu v\right] \cdot \psi \, \mathrm{d}x$$

$$= \iint_\Omega (Qf) \cdot \hat{\psi} \, \mathrm{d}x.$$

**Assumption III.26.** *Let $\Omega \subset \mathbb{R}^3$ be a bounded Lipschitz domain such that the complement $\mathbb{R}^3 \setminus \overline{\Omega}$ is connected. Let $k > 0$ be the wave number and $\varepsilon, \mu \in L^\infty(\Omega, \mathbb{R})$ and $\beta \in L^\infty(\Omega, \mathbb{R})$ real valued. We extend $\varepsilon$ and $\mu$ by one and $\beta$ by zero outside of $\Omega$. We assume:*

*(a) $-Q$ satisfies part (a) and (b) of Assumption III.21.*

*(b) For all $(g, h) \in L^2(\Omega, \mathbb{C}^3) \times L^2(\Omega, \mathbb{C}^3)$ there exists a unique radiating solution of the transmission problem (2.20) for $\kappa = k$.*

*(c) There exists a constant $c > 0$ such that $\tfrac{1}{\varepsilon} - k^2(\mu^2 + \mu)\beta^2 \geq c$ a.e. in $\Omega$.*

$Q$ is negative definite if $q_\mu < 0$, $q_\varepsilon < 0$ and $1 - k^2\mu\beta^2/(q_\varepsilon q_\mu) > 0$. Again part (c) means that $k^2\beta^2$ has to be sufficiently small.

Let $v_0$ be the solution of the slightly different equation

$$\iint_{\mathbb{R}^3} \left[\left(\tfrac{1}{\varepsilon} - k^2\mu\beta^2\right) \mathrm{curl}\, v_0 - k^2\mu\beta v_0\right] \cdot \mathrm{curl}\, \psi - k^2\left[\mu\beta\,\mathrm{curl}\, v_0 - v_0\right] \cdot \psi \, \mathrm{d}x$$

$$\iint_\Omega (Qf) \cdot \hat{\psi} \, \mathrm{d}x. \tag{3.13}$$

Note, that (3.13) differs from the equation for $v_0$ in the case of positive contrasts (cf. (3.9)). Define $\mathcal{T}_0 \colon X \to L^2_{-Q}(\Omega)$ by

$$\mathcal{T}_0 f = -(f + \hat{v}_0). \tag{3.14}$$

Then $\widetilde{\mathcal{T}} f = \mathcal{T}_0 f + (\widetilde{\mathcal{T}} - \mathcal{T}_0)f$ and $(\widetilde{\mathcal{T}} - \mathcal{T}_0)f = \hat{v}_0 - \hat{v}$.

**Proposition III.27.** *Let Assumption III.26 be satisfied.*

*(a) The variational equation (3.13) for $v_0$ is uniquely solvable.*

## 3. Properties of the middle operator

(b) The mapping $f \mapsto v_0|_\Omega$ is compact from $X$ into $L^2_{-Q}(\Omega)$.

*Proof.* (a) The proof is similar to the one for positive contrasts III.23. We only show coercivity of the bilinear form

$$a(w,\psi) := \iint_{\mathbb{R}^3} \left(\tfrac{1}{\varepsilon} - k^2\mu\beta^2\right) \operatorname{curl} w \cdot \operatorname{curl} \overline{\psi} \, dx$$
$$+ k^2 \iint_{\mathbb{R}^3} w \cdot \overline{\psi} - \mu\beta w \cdot \operatorname{curl} \overline{\psi} - \mu\beta \operatorname{curl} w \cdot \overline{\psi} \, dx.$$

$$a(w,w) =$$
$$= \iint_{\mathbb{R}^3} \left(\tfrac{1}{\varepsilon} - k^2(\mu^2+\mu)\beta^2\right) \operatorname{curl} w \cdot \operatorname{curl} \overline{w} + k^2 \mu \beta^2 \operatorname{curl} w \cdot \operatorname{curl} \overline{w} \, dx$$
$$+ k^2 \iint_{\mathbb{R}^3} w \cdot \overline{w} - \mu\beta w \cdot \operatorname{curl} \overline{w} - \mu\beta \operatorname{curl} w \cdot \overline{w} \, dx$$
$$= \iint_{\mathbb{R}^3} \left(\tfrac{1}{\varepsilon} - k^2(\mu^2+\mu)\beta^2\right) |\operatorname{curl} w|^2 + k^2 |\mu\beta \operatorname{curl} w - w|^2 \, dx$$
$$\geq \min\{c, 1\} \|w\|^2_{\mu\beta}.$$

(b) Similar to the proof of Proposition III.23 we can determine a solution $q^n \in H^1(B)$ of the Neumann problem

$$\iint_B \nabla q^n \cdot \nabla \overline{\varphi} + q^n \overline{\varphi} \, dx = \int_{\partial B} (\nu \cdot v_0^n) \overline{\varphi} \, ds$$

for all $\varphi \in H^1(B)$. Then $v^n|_\Omega - \nabla q^n$ satisfies the conditions for the strong convergence by Weber [39] and also $\|\nabla q^n\|_{L^2} \to 0$. $\square$

As in the case of positive constrasts we show coercivity of $\mathcal{T}_0$ (or $-\mathcal{T}_0$ in fact) and compactness of $\widetilde{\mathcal{T}} - \mathcal{T}_0$. Here, the difference $u := v_0 - v$ solves

$$\iint_{\mathbb{R}^3} \left[\left(\tfrac{1}{\varepsilon} - k^2\mu\beta^2\right) \operatorname{curl} u - k^2 \mu \beta u\right] \cdot \operatorname{curl} \psi - k^2 \left[\mu\beta \operatorname{curl} u + \mu u\right] \cdot \psi \, dx$$
$$= -k^2 \iint_{\mathbb{R}^3} (\mu+1) v_0 \cdot \psi \, dx \qquad (3.15)$$

**Theorem III.28.** *Let Assumption III.26 be satisfied and assume that* $1 - k^2\varepsilon\mu\beta^2 \neq 0$ *a.e.*

*(a) The operator $\mathcal{T}_0$ is self–adjoint and $-\mathcal{T}_0$ is coercive on $X$; that is, there exists a constant $c > 0$ such that*

$$-(\mathcal{T}_0 f, f)_{L^2_{-Q}} \geq c\|f\|^2_{L^2_{-Q}} \quad \text{for all } f \in X.$$

*(b) The operator $\widetilde{\mathcal{T}} - \mathcal{T}_0$ is compact from $X$ into $L^2_{-Q}(\Omega)$.*

*Proof.* (a) (i) Analogous to the case of positive contrasts $\mathcal{T}_0$ is self–adjoint. Note that here again the variational equation (3.13) is symmetric in $v_0$ and $\psi$.

(ii) $-\mathcal{T}_0$ is coercive. First we rewrite equation (3.13): Define the function $w := f + \hat{v}_0$ on $\Omega$. Then (3.13) can be written as

$$\iint_{\mathbb{R}^3} \operatorname{curl} v_0 \cdot \operatorname{curl} \psi + k^2\mu v_0 \cdot \psi \, \mathrm{d}x = \iint_{\Omega} (Qw) \cdot \hat{\psi} \, \mathrm{d}x.$$

With $f = w - \hat{v}_0$ we have

$$-(\mathcal{T}_0 f, f)_{L^2_{-Q}} = -(-w, w - \hat{v}_0)_{L^2_{-Q}}$$
$$= (w, w)_{L^2_{-Q}} + (Qw, \hat{v}_0)_{L^2}$$
$$= \|w\|^2_{L^2_{-Q}} + \iint_{\mathbb{R}^3} |\operatorname{curl} v_0|^2 + k^2\mu|v_0|^2 \, \mathrm{d}x$$
$$\geq \|w\|^2_{L^2_{-Q}}.$$

(b) The proof of this part analogous to the case of positive contrasts. □

Figure III.4 shows the different cases for real valued material functions: In the two cases where $q_\varepsilon q_\mu > 0$ we can prove coercivity under additional assumptions. The remaining two cases – here $q_\varepsilon q_\mu < 0$ – are indefinite in the sense that we can not prove coercivity.

## 4. Localization of the scatterer

We want to localize the chiral body; that is, for a given point $z \in \mathbb{R}^3$ we have to decide whether or not it belongs to $\Omega$. We will provide such a

4. Localization of the scatterer

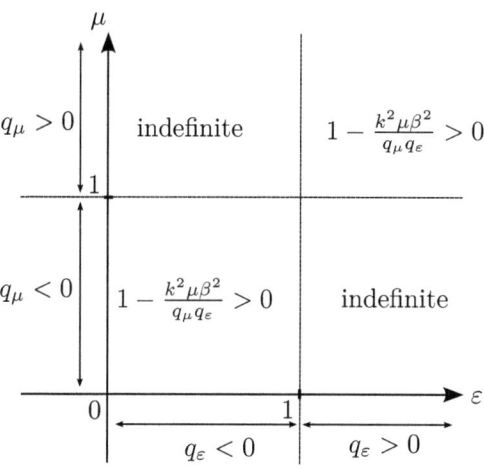

Figure III.4: Real–valued case

decision making possibility by checking whether or not a certain function $\phi_z$ depending on $z$ is in the range of $\mathcal{H}^*$. We start with an integral representation of the exponential function $\theta \mapsto e^{-ik\,\theta \cdot z}$ taken from [24].

**Lemma III.29.** *For $z \in \Omega$ there exists $\widetilde{\varphi} \in L^2(\Omega)$ such that*

$$e^{-ik\,d\cdot z} = \iint_\Omega \widetilde{\varphi}(y)\, e^{-ik\,d\cdot y}\, dy \quad \text{for } d \in \mathbb{S}^2.$$

We adapt this result to our problem. We look for a function $\phi_z$ which can be written as $\mathcal{H}^*\varphi$ for some $\varphi \in L^2(\Omega, \mathbb{C}^3)^2$.

**Theorem III.30.** *For any $z \in \mathbb{R}^3$ and fixed $p \in \mathbb{C}^3 \smallsetminus \{0\}$ define the tangential field $\phi_z \in L^2_t(\mathbb{S}^2)$ by*

$$\phi_z(d) := \big\{(d \times p \times d) + ik\,(d \times p)\big\} e^{-ik\,d\cdot z}, \qquad d \in \mathbb{S}^2. \tag{3.16}$$

*Then $z \in \Omega$ if, and only if, $\phi_z \in \mathcal{R}(\mathcal{H}^*)$.*

*Proof.* By the lemma: For $z \in \Omega$ there exists $\widetilde{\varphi}$ s.t.

$$e^{-ik\,d\cdot z} = \iint_\Omega \widetilde{\varphi}(y)\, e^{-ik\,d\cdot y}\, dy$$

We conclude $\phi_z = \mathcal{H}_1^* \varphi_1 + \mathcal{H}_2^* \varphi_2$ with $\varphi_j = p\widetilde{\varphi}$.
We continue as in the proof of Theorem 4.3 in [20]. Let $z \in \mathbb{R}^3 \setminus \Omega$ and assume, on the contrary, that there exists $\varphi \in L^2(\Omega, \mathbb{C}^3)^2$ such that $\phi_z = \mathcal{H}^* \varphi$. Then $w^\infty = \mathcal{H}^* \varphi$ is the far field pattern of the radiating weak solution to $\operatorname{curl}^2 w - k^2 w = \varphi_1 + \operatorname{curl} \varphi_2$ in $\Omega$ and $\operatorname{curl}^2 w - k^2 w = 0$ in $\mathbb{R}^3 \setminus \overline{\Omega}$. $\phi_z$ is the far field pattern of $\frac{1}{k^2} \operatorname{curl}^2 [p\, \Phi_k(\cdot, z)] + \operatorname{curl}[p\, \Phi_k(\cdot, z)]$. From $\phi_z = \mathcal{H}^* \varphi = w^\infty$ we conclude that

$$w \equiv \frac{1}{k^2} \operatorname{curl}^2 [p\, \Phi_k(\cdot, z)] + \operatorname{curl}[p\, \Phi_k(\cdot, z)] \quad \text{on } \mathbb{R}^3 \setminus (\overline{\Omega} \cup \{z\})$$

by Rellich's Lemma and analytic continuation. But the right–hand side $\frac{1}{k^2} \operatorname{curl}^2 [p\, \Phi_k(\cdot, z)] + \operatorname{curl}[p\, \Phi_k(\cdot, z)]$ has a singularity in $z$, but $w$ is analytic outisde of $\overline{\Omega}$. This a contradiction. □

In the inverse problem the far field data is given; that is, the far field operator $\mathcal{F}$. So we need to characterize the range of $\mathcal{H}^*$ by the range of $\mathcal{F}$ or rather by information that can be extracted from $\mathcal{F}$. Here the factorization in combination with the properties of the middle operator $\mathcal{T}$ gives the main result. We start with absorbing media. The material functions $\varepsilon$ and $\mu$ have non-vanishing imaginary parts.

**Theorem III.31 (Absorbing media).** *Let Assumption III.11 be satisfied and assume that there exist constants* $\gamma_\varepsilon, \gamma_\mu > 0$ *such that* $\operatorname{Im} q_\varepsilon \geq \gamma_\varepsilon$ *and* $\operatorname{Im} q_\mu \geq \gamma_\mu$ *a.e. in* $\Omega$. *Define* $\phi_z$ *by* (3.16) *for* $z \in \mathbb{R}^3$. *Then* $z \in \Omega$ *if, and only if,* $\phi_z \in \mathcal{R}\left((\operatorname{Im} \mathcal{F})^{1/2}\right)$.

*Proof.* The imaginary part of an operator $A - \operatorname{Im} A = \frac{1}{2i}(A - A^*) -$ is self–adjoint. By the factorization of $\mathcal{F}$ the imaginary part $\operatorname{Im} \mathcal{F}$ admits the factorization $\operatorname{Im} \mathcal{F} = \mathcal{H}^*(\operatorname{Im} \mathcal{T})\mathcal{H}$. $\operatorname{Im} \mathcal{F}$ is compact because $\mathcal{H}$ is compact. Furthermore $\left((\operatorname{Im} \mathcal{T})f, f\right)_{L^2} = \operatorname{Im}\left(\mathcal{T}f, f\right)_{L^2}$ and therefore $\operatorname{Im} \mathcal{T}$ is coercive on $\mathcal{R}(\mathcal{H})$ by Theorem III.19. Then, by Corollary 1.22 in [24] the ranges of $\mathcal{H}^*$ and $(\operatorname{Im} \mathcal{F})^{1/2}$ coincide and Theorem III.30 finishes the proof. □

The following conclusions are standard for the Factorization method. From Corollary III.20 we know that $\operatorname{Im} \mathcal{F}$ admits a complete eigensytsem $\{\lambda_j, \psi_j\}_{j \in \mathbb{N}}$ with positive eigenvalues $\lambda_j$. Thus the operator $(\operatorname{Im} \mathcal{F})^{(1/2)}$ has the eigensystem $\{\sqrt{\lambda_j}, \psi_j\}_{j \in \mathbb{N}}$ and $\phi_z$ is an element of $\mathcal{R}\left((\operatorname{Im} \mathcal{F})^{1/2}\right)$

## 4. Localization of the scatterer

if there exists a $p \in L_t^2(\mathbb{S}^2)$ such that $\phi_z = \sum_{j \in \mathbb{N}} \sqrt{\lambda_j}(p, \psi_j)\psi_j$. Then $p$ is formally given by

$$p = \sum_{j \in \mathbb{N}} \frac{(\phi_z, \psi_j)}{\sqrt{\lambda_j}} \psi_j$$

and $p \in L_t^2(\mathbb{S}^2)$ if, and only if,

$$\sum_{j \in \mathbb{N}} \frac{|(\phi_z, \psi_j)|^2}{\lambda_j} < \infty.$$

Or equivalently

$$z \in \Omega \iff W(z) := \left[ \sum_{j \in \mathbb{N}} \frac{|(\phi_z, \psi_j)|^2}{\lambda_j} \right]^{-1} > 0.$$

Now we can give an explicite term for the characteristic function of $\Omega$,

$$\chi_\Omega := \operatorname{sgn} W.$$

In the non-absorbing case we use the modified factorization $\mathcal{F} = \mathcal{H}^\dagger \mathcal{P} \widetilde{\mathcal{T}} \mathcal{H}$ and apply the range identity result from [28], which has been generalized slightly by Sandfort in his PhD thesis [37]:

**Theorem III.32.** *Let $X, Y$ Hilbert spaces, $B: Y \to Y$, $H: Y \to X$ and $A: X \to X$ linear bounded operators with*

$$B = H^* A H.$$

*Assume that*

*(a) $H$ is compact and injective.*

*(b) For some $\alpha \in [0, 2\pi)$ the operator $\operatorname{Re}(e^{i\alpha} A)$ has the form $C + K$ with an coercive and a compact operator $C, K : X \to X$, respectively.*

*(c) $\operatorname{Im} A$ is non-negative on $X$.*

*(d) $A$ is injective.*

*Then the operator $B_\# := |\operatorname{Re}(e^{i\alpha} B)| + \operatorname{Im} B$ is positive and the ranges of $H^* : X \to Y$ and $B_\#^{1/2} : Y \to Y$ coincide.*

**Theorem III.33 (Non–absorbing case).** *Let $k > 0$ the wavenumber. Let the scattering obstacle $\Omega$ and the material parameters – summarized in the matrix $Q$ – fullfill Asumption III.21 or III.26. For $z \in \mathbb{R}^3$ define $\phi_z$ by (3.16). Then $z \in \Omega$ if, and only if, $\phi_z \in \mathcal{R}(\mathcal{F}_\#^{1/2})$ where the operator $\mathcal{F}_\# := |\operatorname{Re} \mathcal{F}| + \operatorname{Im} \mathcal{F}$.*

*Proof.* For the case of positive contrasts. We verify the assumptions of Theorem III.32 for $X$ defined in III.18, $A = \mathcal{P}\widetilde{\mathcal{T}}$ and $H = \mathcal{H}$. The work is all done we just cite the results from the previous section. $\mathcal{H}$ is an integral operator with smooth kernel function, hence compact. $\mathcal{H}$ is injective and $\mathcal{P}\widetilde{\mathcal{T}}$ inherits the properties of $\widetilde{\mathcal{T}}$: For $f \in X$:

$$(\widetilde{\mathcal{T}}f, f) = (\mathcal{P}\widetilde{\mathcal{T}}f, f) + \big((I - \mathcal{P})\widetilde{\mathcal{T}}f, f\big) = (\mathcal{P}\widetilde{\mathcal{T}}f, f)$$

since the image of $I - \mathcal{P}$ is orthogonal to $X \ni f$. We can decompose $\operatorname{Re} \widetilde{\mathcal{T}} = \operatorname{Re} \mathcal{T}_0 + \operatorname{Re}(\widetilde{\mathcal{T}} - \mathcal{T}_0)$ with $\mathcal{T}_0$ defined in (3.10). There exists a constant $c > 0$ such that $\operatorname{Re}(\mathcal{T}_0 f, f) \geq c\|f\|^2$ for $f \in X$ by part (a) of Theorem III.25. Furthermore, we have shown the compactness of $\widetilde{\mathcal{T}} - \mathcal{T}_0$ in part (b). As in the proof for the properties of $\mathcal{T}$ in the absorbing case (Theorem III.19), we compute:

$$\operatorname{Im}(\widetilde{\mathcal{T}}f, f) = \frac{k}{(4\pi)^2} \int_{\mathbb{S}^2} |v^\infty|^2 \, ds \geq 0$$

where $v^\infty$ is the far field pattern of the solution to the variational equation (3.6). Injectivity of $\mathcal{T}$ and therefore of $\widetilde{\mathcal{T}}$ has been discussed in Remark III.17. The case of negative contrasts is proved analogously. In this case we use $\mathcal{T}_0$ defined in (3.14). □

$\mathcal{F}_\#$ is compact, self–adjoint and injective. Hence, the spectral theorem for compact self–adjoint operators yields the existence of a complete eigensystem $\{\lambda_j, \psi_j\}_{j \in \mathbb{N}}$ with strictly positive eigenvalues $\lambda_j$. Thus

$$\mathcal{F}_\#^{1/2} p = \sum_{j \in \mathbb{N}} \sqrt{\lambda_j} (p, \psi_j) \psi_j$$

and we can deduce the characteristic function of the scatterer $\Omega$ as in the case of absorbing media.

4. Localization of the scatterer 79

## The $(\mathcal{F}^*\mathcal{F})^{1/4}$–method

By Theorem III.10 we know that the far field operator $\mathcal{F}$ is normal in the case of non–absorbing media. Furthermore, the scattering operator $\mathcal{S} = I + \frac{ik}{8\pi^2}\mathcal{F}$ is unitary. We want to apply the range identity result in Theorem 1.23 from Kirsch [24]. The assumptions for this theorem are:

- $\widetilde{\mathcal{T}} = \mathcal{T}_0 + K$ with $\mathcal{T}_0$ self–adjoint and coercive on $\mathcal{R}(\mathcal{H})$, $K$ compact,
- $I + ir\mathcal{F}$ unitary for some $r > 0$,
- $\operatorname{Im}(\widetilde{\mathcal{T}}f, f) \neq 0$ for all $f \in \overline{\mathcal{R}(\mathcal{H})}$,
- $\mathcal{F}$ injective.

The first two assumptions have already been chequed. So we study under which conditions the third assumption holds: We already know that

$$\operatorname{Im}(\widetilde{\mathcal{T}}f, f) = \frac{k}{(4\pi)^2} \int_{\mathbb{S}^2} |v^\infty|^2 \, \mathrm{d}s$$

where $v^\infty$ is the far field pattern of the solution to the variational equation (3.6), namely

$$\iint_{\mathbb{R}^3} \left[\left(\tfrac{1}{\varepsilon} - k^2\mu\beta^2\right)\operatorname{curl} v - k^2\mu\beta v\right] \cdot \operatorname{curl}\psi - k^2\left[\mu\beta\operatorname{curl} v + \mu v\right] \cdot \psi \, \mathrm{d}x$$

$$= \iint_\Omega (Qf) \cdot \hat{\psi} \, \mathrm{d}x.$$

for all $\psi \in H_c(\operatorname{curl}, \mathbb{R}^3)$. Hence, $\operatorname{Im}(\widetilde{\mathcal{T}}f, f) = 0$ if, and only if, $v^\infty$ vanishes which implies that $v \equiv 0$ in $\mathbb{R}^3 \setminus \overline{\Omega}$. Then (3.6) reads

$$\iint_\Omega \left[\left(\tfrac{1}{\varepsilon} - k^2\mu\beta^2\right)\operatorname{curl} v - k^2\mu\beta v\right] \cdot \operatorname{curl}\psi - k^2\left[\mu\beta\operatorname{curl} v + \mu v\right] \cdot \psi \, \mathrm{d}x$$

$$= \iint_\Omega (Qf) \cdot \hat{\psi} \, \mathrm{d}x.$$

for all $\psi \in H(\operatorname{curl}, \Omega)$. We conclude that $v \in H(\operatorname{curl}, \Omega)$ with $\nu \times v = 0$ on $\Gamma = \partial\Omega$ and denote by $H_0(\operatorname{curl}, \Omega)$ the space of functions in $H(\operatorname{curl}, \Omega)$ with vanishing trace. This is an interior transmission eigenvalue problem for the parameter $k$:

**Problem 4 (Interior transmission eigenvalue problem).** Given $k > 0$. Determine $v \in H_0(\mathrm{curl}, \Omega)$ and $f \in L_Q^2(\Omega)$ such that

$$\iint_\Omega [(\tfrac{1}{\varepsilon} - k^2\mu\beta^2)\mathrm{curl}\, v - k^2\mu\beta v] \cdot \mathrm{curl}\,\psi - k^2[\mu\beta\,\mathrm{curl}\, v + \mu v] \cdot \psi\, \mathrm{d}x$$

$$= \iint_\Omega (Qf) \cdot \hat{\psi}\, \mathrm{d}x.$$

for all $\psi \in H(\mathrm{curl}, \Omega)$ and

$$\iint_\Omega f \cdot (\mathrm{curl}^2 w - k^2 w)\, \mathrm{d}x = 0$$

for all $w \in H_{00}(\mathrm{curl}^2, \Omega) \times H_{00}(\mathrm{curl}^2, \Omega)$ with $(\mathrm{curl}^2 w - k^2 w) \in L_{Q^{-1}}^2(\Omega)$.

We call $k$ an INTERIOR TRANSMISSION EIGENVALUE if Problem 4 has a non–trivial solution $(v, f) \in H_0(\mathrm{curl}, \Omega) \times L_Q^2(\Omega)$ for $k$.

We conclude that $\mathrm{Im}\,(\widetilde{\mathcal{T}}f, f) \neq 0$ if $k$ is not an interior transmission eigenvalue. Furthermore, $\mathcal{F}$ is injective if $k$ is not an interior transmission eigenvalue: $\mathcal{F}p = 0$ implies $w \equiv 0$ outside of $\Omega$ where $w \in H_{\mathrm{loc}}(\mathrm{curl}, \mathbb{R}^3)$ is the radiating solution of

$$\iint_{\mathbb{R}^3} \left(\tfrac{1}{\varepsilon} - k^2\mu\beta^2\right) \mathrm{curl}\, w \cdot \mathrm{curl}\,\psi - k^2\mu w \cdot \psi\, \mathrm{d}x$$

$$-k^2 \iint_\Omega \mu\beta\, [w \cdot \mathrm{curl}\,\psi + \mathrm{curl}\, w \cdot \psi]\, \mathrm{d}x$$

$$= \iint_\Omega (q_\varepsilon + k^2\mu\beta^2)\, \mathrm{curl}\, w^i \cdot \mathrm{curl}\,\psi + k^2 q_\mu w^i \cdot \psi\, \mathrm{d}x$$

$$+ k^2 \iint_\Omega \mu\beta\, [w^i \cdot \mathrm{curl}\,\psi + \mathrm{curl}\, w^i \cdot \psi]\, \mathrm{d}x$$

for all $\psi \in H_c(\mathrm{curl}, \mathbb{R}^3)$ with $w^i(y) = (\mathcal{H}_1 p)(y) = \int_{\mathbb{S}^2} p(d)e^{ikd \cdot y}\, \mathrm{d}s(d)$ (compare beginning of section 2 in this chapter). Since $w \equiv 0$ outside of $\Omega$ we conclude that $w \in H_0(\mathrm{curl}, \Omega)$ and with $f = (w^i, \mathrm{curl}\, w^i)^\top$ we have a solution to Problem 4. It is the trivial solution because $k$ is not an interior transmission eigenvalue. In particular, $w^i = \mathcal{H}_1 p = 0$ which implies $p = 0$.

Finally, we are able to state the following

## 4. Localization of the scatterer

**Theorem III.34 ($(\mathcal{F}^*\mathcal{F})^{1/4}$–method).** *Assume that the wave number $k > 0$ is not an interior transmission eigenvalue. Let the scattering obstacle $\Omega$ and the material parameters fullfill Asumption III.21 or III.26. For $z \in \mathbb{R}^3$ define $\phi_z$ by (3.16). Then $z \in \Omega$ if, and only if, $\phi_z \in \mathcal{R}((\mathcal{F}^*\mathcal{F})^{1/4})$.*

In this case $\mathcal{F}$ is compact, normal and injective. The spectral theorem for compact normal operators yields the existence of a complete eigensystem $\{\lambda_j, \psi_j\}_{j \in \mathbb{N}}$ with complex eigenvalues $\lambda_j \in \mathbb{C} \smallsetminus \{0\}$ and corresponding normalized eigenfunctions $\psi_j \in L_t^2(\mathbb{S}^2)$. Again, we derive a criterion to determine the scatterer $\Omega$:

$$z \in \Omega \iff \left[\sum_{j \in \mathbb{N}} \frac{|(\phi_z, \psi_j)|^2}{|\lambda_j|}\right]^{-1} > 0.$$

CHAPTER IV

# Scattering by a chiral sphere

In spherical coordinates it is possible to give series expansions for solutions of spherical transmission problems. We study the scattering by a homogeneous chiral ball. All material parameters are real valued. Berketis and Athanasiadis [9] study a similar problem: scattering by perfectly conducting sphere situated in a chiral medium.

In the first section we solve the direct transmission problem. First we recall the main steps for the deduction of vector spherical harmonics. We just give the results and refer to Colton and Kress [15]. Together with the spherical Bessel and Hankel functions they constitute the basic solutions to Maxwell's equations: the vector wave functions. Then we treat the achiral problem: Starting with the series representation of an incident field we give series expansions for the scattered field and the far field pattern depending on the coefficients of the incident field. For the chiral transmission problem we use Bohren's decomposition [16]: The electric and magnetic field are decomposed into a sum of Beltrami fields, which satisfy the achiral Maxwell equations for different wave numbers. Thus, we can directly apply the achiral results to the chiral case.

The second section is devoted to the far field operator. In the spherical case we can express $\mathcal{F}$ explicitly and compute the eigenvalues und eigenfunctions – again, for the achiral and the chiral case. In [24] we find this kind of computation for sound soft scattering. In [20] a spherical achiral transmission problem is considered. With the aid of the eigensystem we can evaluate the series which is used for the characteristic function of the scatterer in Theorem III.34 and determine the scatterer explicitly.

## 1. Spherical transmission problems

The first part is a brief summary of the sections 2.3, 2.4 and 6.5 in [15].

### 1.1. Spherical vector wave functions

We are looking for solutions of Maxwell's equations in spherical coordinates. It is possible to construct such solutions — spherical vector wave functions — from solutions of the Helmholtz equation. In spherical coordinates $(\rho, \theta, \varphi)$ with

$$x = (\rho \sin \theta \cos \varphi, \rho \sin \theta \sin \varphi, \rho \cos \theta)^\top \in \mathbb{R}^3,$$

where $\rho \geq 0$, $\theta \in [0, \pi]$, $\varphi \in [0, 2\pi]$, the Helmholtz equation takes the form

$$\frac{1}{\rho^2} \frac{\partial}{\partial \rho} \left( \rho^2 \frac{\partial u}{\partial \rho} \right) + \frac{1}{\rho^2 \sin \theta} \frac{\partial}{\partial \theta} \left( \sin \theta \frac{\partial u}{\partial \theta} \right) + \frac{1}{\rho^2 \sin^2 \theta} \frac{\partial^2 u}{\partial \varphi^2} + k^2 u = 0.$$

Separation of variables

$$u(\rho, \theta, \varphi) = u_1(\rho) u_2(\theta, \varphi)$$

leads to spherical harmonics and spherical Bessel functions. The spherical harmonics are given by

$$Y_n^m(\theta, \varphi) := \sqrt{\frac{2n+1}{4\pi} \frac{(n-|m|)!}{(n+|m|)!}} P_n^{|m|}(\cos \theta) e^{im\varphi}$$

for $m = -n, \ldots, n$ and $n = 0, 1, 2, \ldots$ Here, $P_n^m$ denotes the associated Legendre polynomial

$$P_n^m(t) := (1-t^2)^{m/2} \frac{d^m P_n(t)}{dt^m}, \qquad m = 0, \ldots, n,$$

which solves the associated Legendre differential equation

$$(1-t^2) f''(t) - 2t f'(t) + \left\{ n(n+1) - \frac{m^2}{1-t^2} \right\} f(t) = 0.$$

## 1. Spherical transmission problems

$P_n$ is the Legendre polynomial which satisfies the Legendre differential equation

$$(1-t^2)P_n''(t) - 2tP_n'(t) + n(n+1)P_n(t) = 0 \qquad n = 0, 1, 2, \ldots$$

The radial part of the Helmholtz equation is given by the spherical Bessel differential equation

$$t^2 f''(t) + 2t f'(t) + \left[t^2 - n(n+1)\right] f(t) = 0$$

which is satisfied by spherical Bessel and Neumann functions $j_n$ and $y_n$, respectively, for $n = 0, 1, 2, \ldots$

$$j_n(t) := \sum_{p=0}^{\infty} \frac{(-1)^p t^{n+2p}}{2^p p! \, 1 \cdot 3 \cdots (2n+2p+1)},$$

$$y_n(t) := -\frac{(2n)!}{2^n n!} \sum_{p=0}^{\infty} \frac{(-1)^p t^{2p-n-1}}{2^p p! (-2n+1)(-2n+3) \cdots (-2n+2p-1)}.$$

Linear combination gives the spherical Hankel function of the first kind $h_n = h_n^{(1)}$ with

$$h_n := j_n + i\, y_n, \qquad n = 0, 1, 2, \ldots$$

Finally, the following functions are solutions of the Helmholtz equation in spherical coordinates: For $n \in \mathbb{N}_0$ and $-n \le m \le n$

$$u_n^m(x) = j_n(k|x|) Y_n^m(\hat{x})$$

is an entire solution to the Helmholtz equation and

$$v_n^m(x) = h_n(k|x|) Y_n^m(\hat{x})$$

is a radiating solution to the Helmholtz equation in $\mathbb{R}^3 \setminus \{0\}$.

We use these to construct such solutions for Maxwell's equations

$$\operatorname{curl} E = ikH \quad \text{and} \quad \operatorname{curl} H = -ikE.$$

For $n \in \mathbb{N}_0$ and $-n \le m \le n$ the functions

$$M_n^m(x) := \frac{1}{\sqrt{n(n+1)}} \operatorname{curl}\left[x\, u_n^m(x)\right], \qquad \frac{1}{ik} \operatorname{curl} M_n^m(x)$$

are an entire solution to Maxwell's equations and

$$N_n^m(x) := \frac{1}{\sqrt{n(n+1)}} \operatorname{curl}\left[x\, v_n^m(x)\right], \qquad \frac{1}{ik} \operatorname{curl} N_n^m(x)$$

are a radiating solution to Maxwell's equations in $\mathbb{R}^3 \smallsetminus \{0\}$.

Define the vector spherical harmonics $U_n^m$ and $V_n^m$ for $n = 0, 1, 2, \ldots$ and $m = -n, \ldots, n$ by

$$U_n^m(\hat{x}) := \frac{1}{\sqrt{n(n+1)}} \operatorname{Grad} Y_n^m(\hat{x}), \quad V_n^m(\hat{x}) := \hat{x} \times U_n^m(\hat{x}) \quad \text{for } \hat{x} \in \mathbb{S}^2$$

with the surface gradient Grad. $U_n^m$ and $V_n^m$ are tangential fields on the unit sphere. Hence,

$$M_n^m(x) = -j_n(k|x|)V_n^m(\hat{x}),$$
$$N_n^m(x) = -h_n(k|x|)V_n^m(\hat{x})$$

and

$$\operatorname{curl} M_n^m(x) = \frac{1}{|x|}\Big[j_n(k|x|) + k|x|j_n'(k|x|)\Big]U_n^m(\hat{x}),$$

$$\operatorname{curl} N_n^m(x) = \frac{1}{|x|}\Big[h_n(k|x|) + k|x|h_n'(k|x|)\Big]U_n^m(\hat{x}).$$

The tangential traces are given by

$$\hat{x} \times M_n^m(x) = j_n(k|x|)U_n^m(\hat{x}), \tag{4.1}$$
$$\hat{x} \times N_n^m(x) = h_n(k|x|)U_n^m(\hat{x}) \tag{4.2}$$

and

$$\hat{x} \times \operatorname{curl} M_n^m(x) = \frac{1}{|x|}\Big[j_n(k|x|) + k|x|j_n'(k|x|)\Big]V_n^m(\hat{x}), \tag{4.3}$$

$$\hat{x} \times \operatorname{curl} N_n^m(x) = \frac{1}{|x|}\Big[h_n(k|x|) + k|x|h_n'(k|x|)\Big]V_n^m(\hat{x}) \tag{4.4}$$

Finally, we have the following representation for the far field pattern: Let $H^s$ be a radiating solution to Maxwell's equations given as series

$$H^s = \sum a_n^m N_n^m + b_n^m \tfrac{1}{ik} \operatorname{curl} N_n^m.$$

1. Spherical transmission problems

The far field pattern is given by

$$H^\infty = \frac{4\pi}{k} \sum \frac{1}{i^{n+1}} \left[ b_n^m U_n^m - a_n^m V_n^m \right]$$

and satisfies

$$H^s(x) = \frac{e^{ik|x|}}{4\pi|x|} H^\infty(\hat{x}) + \mathcal{O}(|x|^{-2}), \qquad |x| \to \infty.$$

Here and throughout this chapter, we abbreviate

$$\sum s_n^m := \sum_{n=0}^{\infty} \sum_{m=-n}^{n} s_n^m.$$

## 1.2. Spherical Maxwell transmission problem

We start with the setting of the transmission problem. The penetrable scattering obstacle is a ball $B = B(0,1)$ with radius 1 located at the origin. In the exterior there is vacuum. The ball consists of constant lossless material with $\varepsilon \neq \varepsilon_0$ or $\mu \neq \mu_0$. The wavenumber is given by

$$\begin{cases} k = \omega\sqrt{\varepsilon_0\mu_0}, & \text{in } \overline{B}^c, \\ \kappa = \omega\sqrt{\varepsilon\mu}, & \text{in } B. \end{cases}$$

The ball is illuminated by an incident field $H^i$ which is a solution of Maxwell's equations in the exterior of $B$.

$$\operatorname{curl}^2 H^i - k^2 H^i = 0 \quad \text{in } \overline{B}^c$$

The total field $H$ inside the ball satisfies Maxwell's equations for the wavenumber $\kappa$,

$$\operatorname{curl}^2 H - \kappa^2 H = 0 \quad \text{in } B.$$

The scattered field $H^s$ in the exterior of $B$ satisfies Maxwell's equations for the wavenumber $k$ and the Silver–Müller radiation condition.

$$\operatorname{curl}^2 H^s - k^2 H^s = 0 \quad \text{in } \overline{B}^c, \quad \text{radiating}.$$

On the boundary $\{|x| = 1\}$ the transmission conditions are given by the continuity of the tangential traces ($\hat{x} \in \mathbb{S}^2$):

$$\hat{x} \times H^i(\hat{x}) + \hat{x} \times H^s(\hat{x}) = \hat{x} \times H(\hat{x}), \tag{4.5}$$

$$\frac{1}{k}\hat{x} \times \operatorname{curl} H^i(\hat{x}) + \frac{1}{k}\hat{x} \times \operatorname{curl} H^s(\hat{x}) = \frac{1}{\kappa}\hat{x} \times \operatorname{curl} H(\hat{x}). \tag{4.6}$$

In this setting we can expand the field in series of vector wave functions,

$$H^i(x) = \sum \alpha_n^m M_n^m(x,k) + \beta_n^m \frac{1}{ik} \operatorname{curl} M_n^m(x,k) \qquad \text{in } \overline{B}^c,$$

$$H^s(x) = \sum c_n^m N_n^m(x,k) + d_n^m \frac{1}{ik} \operatorname{curl} N_n^m(x,k) \qquad \text{in } \overline{B}^c,$$

$$H(x) = \sum a_n^m M_n^m(x,\kappa) + b_n^m \frac{1}{i\kappa} \operatorname{curl} M_n^m(x,\kappa) \qquad \text{in } B$$

and deduce systems of linear equations to determine the coefficients $a_n^m$, $c_n^m$ and $b_n^m$, $d_n^m$ for $n \in \mathbb{N}_0$ and $m = -n, \ldots, n$. We compute the tangential traces of the series with the aid of the tangential traces for the vector wave functions (4.1)–(4.4).

$$\hat{x} \times H^i(\hat{x}) = \sum \alpha_n^m j_n(k) U_n^m(\hat{x}) + \frac{1}{ik}\beta_n^m \left[j_n(k) + kj_n'(k)\right] V_n^m(\hat{x}),$$

$$\hat{x} \times H^s(\hat{x}) = \sum c_n^m h_n(k) U_n^m(\hat{x}) + \frac{1}{ik}d_n^m \left[h_n(k) + kh_n'(k)\right] V_n^m(\hat{x}),$$

$$\hat{x} \times H(\hat{x}) = \sum a_n^m j_n(\kappa) U_n^m(\hat{x}) + \frac{1}{i\kappa}b_n^m \left[j_n(\kappa) + \kappa j_n'(\kappa)\right] V_n^m(\hat{x})$$

and

$$\frac{1}{k}\hat{x} \times \operatorname{curl} H^i(\hat{x}) = \sum \alpha_n^m \frac{1}{k}\left[j_n(k) + kj_n'(k)\right] V_n^m(\hat{x}) + \beta_n^m \frac{1}{i} j_n(k) U_n^m(\hat{x}),$$

$$\frac{1}{k}\hat{x} \times \operatorname{curl} H^s(\hat{x}) = \sum c_n^m \frac{1}{k}\left[h_n(k) + kh_n'(k)\right] V_n^m(\hat{x}) + d_n^m \frac{1}{i} h_n(k) U_n^m(\hat{x}),$$

$$\frac{1}{\kappa}\hat{x} \times \operatorname{curl} H(\hat{x}) = \sum a_n^m \frac{1}{\kappa}\left[j_n(\kappa) + \kappa j_n'(\kappa)\right] V_n^m(\hat{x}) + b_n^m \frac{1}{i} j_n(\kappa) U_n^m(\hat{x}).$$

Plugging this into the transmission conditions (4.5),(4.6) yields

$$c_n^m h_n(k) + \alpha_n^m j_n(k) \stackrel{!}{=} a_n^m j_n(\kappa),$$

$$d_n^m \frac{1}{ik}\left[h_n(k) + kh_n'(k)\right] + \beta_n^m \frac{1}{ik}\left[j_n(k) + kj_n'(k)\right] \stackrel{!}{=} b_n^m \frac{1}{i\kappa}\left[j_n(\kappa) + \kappa j_n'(\kappa)\right]$$

# 1. Spherical transmission problems

and

$$c_n^m \frac{1}{k}[h_n(k) + kh'_n(k)] + \alpha_n^m \frac{1}{k}[j_n(k) + kj'_n(k)] \stackrel{!}{=} a_n^m \frac{1}{\kappa}[j_n(\kappa) + \kappa j'_n(\kappa)],$$

$$d_n^m \frac{1}{i} h_n(k) + \beta_n^m \frac{1}{i} j_n(k) \stackrel{!}{=} b_n^m \frac{1}{i} j_n(\kappa)$$

for $n \in \mathbb{N}_0$ and $m = -n, \ldots, n$. The two resulting linear systems can be summarized as

$$\begin{pmatrix} j_n(\kappa) & -h_n(k) \\ \frac{1}{\kappa} j_n(\kappa) + j'_n(\kappa) & -\frac{1}{k} h_n(k) - h'_n(k) \end{pmatrix} \begin{pmatrix} a_n^m & b_n^m \\ c_n^m & d_n^m \end{pmatrix}$$

$$= \begin{pmatrix} j_n(k) \\ \frac{1}{k} j_n(k) + j'_n(k) \end{pmatrix} \begin{pmatrix} \alpha_n^m & \beta_n^m \end{pmatrix}$$

with determinant

$$\det{}_n(\kappa) = \left(\frac{1}{\kappa} - \frac{1}{k}\right) h_n(k) j_n(\kappa) + h_n(k) j'_n(\kappa) - h'_n(k) j_n(\kappa)$$

and inverse

$$\frac{1}{\det{}_n(\kappa)} \begin{pmatrix} -\frac{1}{k} h_n(k) - h'_n(k) & h_n(k) \\ -\frac{1}{\kappa} j_n(\kappa) - j'_n(\kappa) & j_n(\kappa) \end{pmatrix}.$$

Hence, the solutions are given by

$$\begin{pmatrix} a_n^m & b_n^m \\ c_n^m & d_n^m \end{pmatrix} = -\frac{1}{\det{}_n(\kappa)} \begin{pmatrix} \frac{i}{k^2} \\ \operatorname{Re} \det{}_n(\kappa) \end{pmatrix} \begin{pmatrix} \alpha_n^m & \beta_n^m \end{pmatrix}$$

**Proposition IV.1.** $\det{}_n(\kappa) \neq 0$.

*Proof.* Assume to the contrary $\det{}_n(\kappa) = 0$; that is,

$$\det{}_n(\kappa) = \left(\frac{1}{\kappa} - \frac{1}{k}\right) h_n(k) j_n(\kappa) + h_n(k) j'_n(\kappa) - h'_n(k) j_n(\kappa)$$

$$= \left(\frac{1}{\kappa} - \frac{1}{k}\right) j_n(k) j_n(\kappa) + j_n(k) j'_n(\kappa) - j'_n(k) j_n(\kappa)$$

$$+ i \left[\left(\frac{1}{\kappa} - \frac{1}{k}\right) y_n(k) j_n(\kappa) + y_n(k) j'_n(\kappa) - y'_n(k) j_n(\kappa)\right]$$

$$\stackrel{!}{=} 0$$

$$\left(\frac{1}{\kappa}-\frac{1}{k}\right)+\frac{j'_n(\kappa)}{j_n(\kappa)}=\frac{j'_n(k)}{j_n(k)} \quad \text{and} \quad \left(\frac{1}{\kappa}-\frac{1}{k}\right)+\frac{j'_n(\kappa)}{j_n(\kappa)}=\frac{y'_n(k)}{y_n(k)}$$

$\Longrightarrow 0 \stackrel{!}{=} y'_n(k)j_n(k) - j'_n(k)y_n(k) = 1/k^2$ (Wronskian)
Contradiction! □

**Summary**

Given the incident field

$$H^i(x) = \sum \alpha_n^m M_n^m(x,k) + \beta_n^m \frac{1}{ik} \operatorname{curl} M_n^m(x,k),$$

the total field inside the scatterer $B$ is given by

$$H(x) = -\frac{i}{k^2} \sum \frac{1}{\det_n(\kappa)} \left[\alpha_n^m M_n^m(x,\kappa) + \frac{1}{i\kappa}\beta_n^m \operatorname{curl} M_n^m(x,\kappa)\right]$$

and the (radiating) scattered field outside the scatterer is given by

$$H^s(x) = -\sum \frac{\operatorname{Re}\det_n(\kappa)}{\det_n(\kappa)} \left[\alpha_n^m N_n^m(x,k) + \frac{1}{ik}\beta_n^m \operatorname{curl} N_n^m(x,k)\right].$$

The far field pattern is given by

$$H^\infty(\hat{x}) = \frac{4\pi}{k} \sum \frac{1}{i^{n+1}} \frac{\operatorname{Re}\det_n(\kappa)}{\det_n(\kappa)} \left[\alpha_n^m V_n^m(\hat{x}) - \beta_n^m U_n^m(\hat{x})\right].$$

We apply these results to Beltrami fields which appear in our chiral transmission problem.

## 1.3. Spherical chiral transmission problem

The setting is similar to the one in the previous subsection. But now, the medium inside the ball is (homogeneous, lossless and) chiral. More precisely, let the permittivity, permeability and chirality be defined by

$$\varepsilon = \begin{cases} \varepsilon_0 & \text{in } \overline{B}^c, \\ \varepsilon_B & \text{in } B, \end{cases} \quad \mu = \begin{cases} \mu_0 & \text{in } \overline{B}^c, \\ \mu_B & \text{in } B, \end{cases} \quad \beta = \begin{cases} 0 & \text{in } \overline{B}^c, \\ \beta_B & \text{in } B. \end{cases}$$

# 1. Spherical transmission problems

with real constants $\varepsilon_B, \mu_B, \beta_B$. Define the wave number by

$$\begin{cases} k := \omega\sqrt{\varepsilon_0\mu_0} & \text{in } \overline{B}^c, \\ \kappa := \omega\sqrt{\varepsilon_B\mu_B} & \text{in } B. \end{cases}$$

Then the chiral Maxwell's equations read (confere Stratis et al. [6])

$$\operatorname{curl}^2 U - 2\frac{\kappa^2}{1-\kappa^2\beta^2}\beta \operatorname{curl} U - \frac{\kappa^2}{1-\kappa^2\beta^2}U = 0 \quad \text{in } B$$

for $U = E$ or $U = H$ and the fields appearing in our transmission problem satisfy the following equations:

$$\operatorname{curl}^2 U^i - k^2 U^i = 0 \quad \text{in } \overline{B}^c,$$

$$\operatorname{curl}^2 U^s - k^2 U^s = 0 \quad \text{in } \overline{B}^c, \text{ radiating},$$

$$\operatorname{curl}^2 U - 2\frac{\kappa^2}{1-\kappa^2\beta^2}\beta \operatorname{curl} U - \frac{\kappa^2}{1-\kappa^2\beta^2}U = 0 \quad \text{in } B$$

for $U = E$ or $U = H$.

For homogeneous materials it is possible to decompose the electric and magnetic field and use the results from the achiral transmission problem. Compare Athanasiadis, Martin and Stratis [6]. It is Bohren's decomposition [16]. Define $\kappa_L, \kappa_R$ by

$$\kappa_L := \begin{cases} \frac{\kappa}{1-\kappa\beta} & \text{in } B, \\ k & \text{in } \overline{B}^c \end{cases} \quad \text{and} \quad \kappa_R := \begin{cases} \frac{\kappa}{1+\kappa\beta} & \text{in } B, \\ k & \text{in } \overline{B}^c. \end{cases}$$

Then:

$$\kappa_L \kappa_R = \frac{\kappa^2}{1-\kappa^2\beta^2} \quad \text{in } B.$$

Define: $Q_L := E + iH$ and $Q_R := E - iH$. Then $Q_L, Q_R$ are BELTRAMI FIELDS; that is,

$$\operatorname{curl} Q_L = \kappa_L Q_L \quad \text{and} \quad \operatorname{curl} Q_R = -\kappa_R Q_R$$

and they satisfy the achiral Maxwell equations for the wave number $\kappa_L$ and $\kappa_R$, respectively:

$$\operatorname{curl}^2 Q_L - \kappa_L{}^2 Q_L = 0 \quad \text{and} \quad \operatorname{curl}^2 Q_R - \kappa_R{}^2 Q_R = 0.$$

Thus, we can apply the result of the previous subsection to the fields $Q_L$ and $Q_R$ and deduce series representations for the fields $E$ and $H$ since

$$E = \frac{1}{2}(Q_L + Q_R) \quad \text{and} \quad H = \frac{1}{2i}(Q_L - Q_R).$$

We start with the incident fields $H^i$ and $E^i$:

$$H^i(x) = \sum \alpha_n^m M_n^m(x,k) + \beta_n^m \frac{1}{ik} \operatorname{curl} M_n^m(x,k),$$
$$E^i(x) = -\frac{1}{ik} \operatorname{curl} H^i(x) = \sum \beta_n^m M_n^m(x,k) - \alpha_n^m \frac{1}{ik} \operatorname{curl} M_n^m(x,k).$$

We introduce the incident fields $Q_L^i := E^i + iH^i$ and $Q_R^i := E^i - iH^i$:

$$Q_L^i(x) = \sum (\beta_n^m + i\alpha_n^m) M_n^m(x,k) - (\alpha_n^m - i\beta_n^m)\frac{1}{ik} \operatorname{curl} M_n^m(x,k),$$
$$Q_R^i(x) = \sum (\beta_n^m - i\alpha_n^m) M_n^m(x,k) - (\alpha_n^m + i\beta_n^m)\frac{1}{ik} \operatorname{curl} M_n^m(x,k).$$

The total fields $Q_L, Q_R$ indside $B$ are given by

$$Q_L(x) = -\frac{i}{k^2} \sum \frac{1}{\det_n(\kappa_L)} (\beta_n^m + i\alpha_n^m) M_n^m(x,\kappa_L)$$
$$- \frac{1}{i\kappa_L \det_n(\kappa_L)} (\alpha_n^m - i\beta_n^m) \operatorname{curl} M_n^m(x,\kappa_L),$$
$$Q_R(x) = -\frac{i}{k^2} \sum \frac{1}{\det_n(\kappa_R)} (\beta_n^m - i\alpha_n^m) M_n^m(x,\kappa_R)$$
$$- \frac{1}{i\kappa_R \det_n(\kappa_R)} (\alpha_n^m + i\beta_n^m) \operatorname{curl} M_n^m(x,\kappa_R)$$

and the total fields $E$ and $H$ can be computed by $E(x) = \frac{1}{2}(Q_L + Q_R)$ and $H(x) = \frac{1}{2i}(Q_L - Q_R)$. At this point we can see that the fields $E$ and $H$ admit an expansion in terms of the vector wave functions for the wave numbers $\kappa_L$ and $\kappa_R$.

1. Spherical transmission problems

The scattered fields $Q_L^s, Q_R^s$ have the series expansions

$$Q_L^s(x) = -\sum \frac{\operatorname{Re} \det_n(\kappa_L)}{\det_n(\kappa_L)}(\beta_n^m + i\alpha_n^m)N_n^m(x,k)$$
$$\qquad - \frac{\operatorname{Re} \det_n(\kappa_L)}{ik\det_n(\kappa_L)}(\alpha_n^m - i\beta_n^m)\operatorname{curl} N_n^m(x,k),$$

$$Q_R^s(x) = -\sum \frac{\operatorname{Re} \det_n(\kappa_R)}{\det_n(\kappa_R)}(\beta_n^m - i\alpha_n^m)N_n^m(x,k)$$
$$\qquad - \frac{\operatorname{Re} \det_n(\kappa_R)}{ik\det_n(\kappa_R)}(\alpha_n^m + i\beta_n^m)\operatorname{curl} N_n^m(x,k).$$

That is,

$$Q_L^s(x) = -\sum c_L(\beta_n^m + i\alpha_n^m)N_n^m(x,k) - \frac{c_L}{ik}(\alpha_n^m - i\beta_n^m)\operatorname{curl} N_n^m(x,k),$$
$$Q_R^s(x) = -\sum c_R(\beta_n^m - i\alpha_n^m)N_n^m(x,k) - \frac{c_R}{ik}(\alpha_n^m + i\beta_n^m)\operatorname{curl} N_n^m(x,k)$$

where we abbreviate $c_L := \frac{\operatorname{Re} \det_n(\kappa_L)}{\det_n(\kappa_L)}$ and $c_R := \frac{\operatorname{Re} \det_n(\kappa_R)}{\det_n(\kappa_R)}$. Hence, the series expansions for the electric and magnetic field read

$$E^s(x) = -\frac{1}{2}\sum \left[(c_L + c_R)\beta_n^m + (c_L - c_R)i\alpha_n^m\right]N_n^m(x,k)$$
$$\qquad - \frac{1}{ik}\left[(c_L + c_R)\alpha_n^m - (c_L - c_R)i\beta_n^m\right]\operatorname{curl} N_n^m(x,k),$$
$$H^s(x) = -\frac{1}{2i}\sum \left[(c_L - c_R)\beta_n^m + (c_L + c_R)i\alpha_n^m\right]N_n^m(x,k)$$
$$\qquad - \frac{1}{ik}\left[(c_L - c_R)\alpha_n^m - (c_L + c_R)i\beta_n^m\right]\operatorname{curl} N_n^m(x,k).$$

Finally, the far field patterns are given by

$$Q_L^\infty(\hat{x}) = \frac{4\pi}{k}\sum \frac{c_L}{i^{n+1}}\left[(\beta_n^m + i\alpha_n^m)V_n^m(\hat{x}) + (\alpha_n^m - i\beta_n^m)U_n^m(\hat{x})\right],$$
$$Q_R^\infty(\hat{x}) = \frac{4\pi}{k}\sum \frac{c_R}{i^{n+1}}\left[(\beta_n^m - i\alpha_n^m)V_n^m(\hat{x}) + (\alpha_n^m + i\beta_n^m)U_n^m(\hat{x})\right]$$

and

$$E^\infty(\hat{x}) = \frac{4\pi}{2k} \sum \frac{c_L}{i^{n+1}} \left[(\beta_n^m + i\alpha_n^m)V_n^m(\hat{x}) + (\alpha_n^m - i\beta_n^m)U_n^m(\hat{x})\right]$$
$$+ \frac{c_R}{i^{n+1}} \left[(\beta_n^m - i\alpha_n^m)V_n^m(\hat{x}) + (\alpha_n^m + i\beta_n^m)U_n^m(\hat{x})\right],$$
$$H^\infty(\hat{x}) = \frac{4\pi}{2ik} \sum \frac{c_L}{i^{n+1}} \left[(\beta_n^m + i\alpha_n^m)V_n^m(\hat{x}) + (\alpha_n^m - i\beta_n^m)U_n^m(\hat{x})\right]$$
$$- \frac{c_R}{i^{n+1}} \left[(\beta_n^m - i\alpha_n^m)V_n^m(\hat{x}) + (\alpha_n^m + i\beta_n^m)U_n^m(\hat{x})\right].$$

## 2. The far fiel operator

In this section we develop an explicit form of the far field operator $\mathcal{F}$ for the spherical case.

The setting is the same as in the previous subsection. The scattering obstacle is a ball $B = B(0,1)$ filled with homogeneous chiral material situated in vacuum. We already deduced a series representation of the far field pattern $H^\infty$ for a given incident field $H^i$.

$\mathcal{F}$ gives a superposition of the far field patterns $\mathrm{H}^\infty(\hat{x}; d, p)$ induced by plane waves $\mathrm{H}^i(x) = p\, e^{ik\, d \cdot x}$. We start with the series expansion of plane waves. Once we know these coefficients the representation of the far field pattern found in the previous section directly yields the explicit form of $\mathcal{F}$. Again we treat the achiral and the chiral case.

### 2.1. Series expansion of a plane wave

At first, we have to expand a plane wave into a series of the vector wave functions $M_n^m$ and $\operatorname{curl} M_n^m$; that is, we must determine the coefficients $a_n^m$ and $b_n^m$ in the series

$$p e^{ik\, d \cdot x} =$$
$$= \sum a_n^m M_n^m(x, k) + b_n^m \tfrac{1}{ik} \operatorname{curl} M_n^m(x, k)$$
$$= \sum a_n^m (-1) j_n(k|x|) V_n^m(\hat{x}) + b_n^m \tfrac{1}{ik} \left[\tfrac{1}{|x|} j_n(k|x|) + k j_n'(k|x|)\right] U_n^m(\hat{x})$$

## 2. The far fiel operator

with

$$-a_n^m j_n(k|x|) = \int_{\mathbb{S}^2} e^{ik|x|\hat{y}\cdot d}(d\times p\times d)\cdot \overline{V_n^m(\hat{y})}\,\mathrm{d}s(\hat{y}),$$

$$\frac{1}{ik}b_n^m\Big[\frac{1}{|x|}j_n(k|x|)+kj_n'(k|x|)\Big] = \int_{\mathbb{S}^2} e^{ik|x|\hat{y}\cdot d}(d\times p\times d)\cdot \overline{U_n^m(\hat{y})}\,\mathrm{d}s(\hat{y})$$

where we used $p = d \times p \times d$ since $p\cdot d = 0$ and $|d|=1$.

In what follows, we compute the Fourier coefficients on the right hand side. More precisely, we compute the complex conjugate:

$$\int_{\mathbb{S}^2} e^{-ik|x|\hat{y}\cdot d}(d\times \bar{p}\times d)\cdot V_n^m(\hat{y})\,\mathrm{d}s(\hat{y}),$$

$$\int_{\mathbb{S}^2} e^{-ik|x|\hat{y}\cdot d}(d\times \bar{p}\times d)\cdot U_n^m(\hat{y})\,\mathrm{d}s(\hat{y})$$

Here we recognize the far field pattern of $\mathrm{curl}^2[\bar{p}\,\Phi_k(z,|x|\hat{y})]$ with respect to $z = |z|d$. We work with the Stratton–Chu formulae applied on a ball of radius $R$. Therefore we introduce the operators $\mathcal{C}_1$ and $\mathcal{C}_2$ defined for tangential vector fields $\varphi$:

$$(\mathcal{C}_1\varphi)(x) := \mathrm{curl}\int_{|x|=R}\varphi(y)\,\Phi_k(x,y)\,\mathrm{d}s(y),$$

$$(\mathcal{C}_2\varphi)(x) := \mathrm{curl}\,\mathrm{curl}\int_{|x|=R}\varphi(y)\,\Phi_k(x,y)\,\mathrm{d}s(y).$$

Then the Stratton–Chu formulae III.2 and III.3 read

$$-\mathcal{C}_1(\nu\times E)+\tfrac{1}{ik}\mathcal{C}_2(\nu\times H) = \begin{cases} E & \text{in } B(0,R),\\ 0 & \text{in } \overline{B(0,R)}^c, \end{cases}$$

$$-\mathcal{C}_1(\nu\times H)-\tfrac{1}{ik}\mathcal{C}_2(\nu\times E) = \begin{cases} H & \text{in } B(0,R),\\ 0 & \text{in } \overline{B(0,R)}^c, \end{cases}$$

$$\mathcal{C}_1(\nu\times E^s)-\tfrac{1}{ik}\mathcal{C}_2(\nu\times H^s) = \begin{cases} E^s & \text{in } \overline{B(0,R)}^c,\\ 0 & \text{in } B(0,R), \end{cases}$$

$$\mathcal{C}_1(\nu\times H^s)+\tfrac{1}{ik}\mathcal{C}_2(\nu\times E^s) = \begin{cases} H^s & \text{in } \overline{B(0,R)}^c,\\ 0 & \text{in } B(0,R). \end{cases}$$

We apply the interior Stratton–Chu formulae to $M_n^m$, $\frac{1}{ik}\operatorname{curl} M_n^m$ and the exterior formulae to $N_n^m$, $\frac{1}{ik}\operatorname{curl} N_n^m$ for $x \in \overline{B(0,R)}^c$.

$$\tfrac{1}{k^2}\mathcal{C}_2(\nu \times \operatorname{curl} M_n^m) + \mathcal{C}_1(\nu \times M_n^m) = 0,$$
$$\tfrac{1}{k^2}\mathcal{C}_2(\nu \times \operatorname{curl} N_n^m) + \mathcal{C}_1(\nu \times N_n^m) = N_n^m,$$
$$\mathcal{C}_2(\nu \times M_n^m) + \mathcal{C}_1(\nu \times \operatorname{curl} M_n^m) = 0,$$
$$\mathcal{C}_2(\nu \times N_n^m) + \mathcal{C}_1(\nu \times \operatorname{curl} N_n^m) = \operatorname{curl} N_n^m.$$

That is, using the expressions for the tangential traces (4.1)–(4.4) found in the preliminary subsection and abbreviating $j_n = j_n(kR)$, $h_n = h_n(kR)$, $y_n = y_n(kR), \ldots$

$$\tfrac{1}{k^2}\mathcal{C}_2\bigl[(\tfrac{1}{R}j_n + kj_n')V_n^m\bigr] + \mathcal{C}_1[j_n U_n^m] = 0,$$
$$\tfrac{1}{k^2}\mathcal{C}_2\bigl[(\tfrac{1}{R}h_n + kh_n')V_n^m\bigr] + \mathcal{C}_1[h_n U_n^m] = N_n^m,$$
$$\mathcal{C}_1\bigl[(\tfrac{1}{R}j_n + kj_n')V_n^m\bigr] + \mathcal{C}_2[j_n U_n^m] = 0,$$
$$\mathcal{C}_1\bigl[(\tfrac{1}{R}h_n + kh_n')V_n^m\bigr] + \mathcal{C}_2[h_n U_n^m] = \operatorname{curl} N_n^m.$$

Using $h_n = j_n + iy_n$ we conclude

$$\tfrac{1}{k^2}\mathcal{C}_2\bigl[(\tfrac{1}{R}j_n + kj_n')V_n^m\bigr] + \mathcal{C}_1[j_n U_n^m] = 0,$$
$$\tfrac{i}{k^2}\mathcal{C}_2\bigl[(\tfrac{1}{R}y_n + ky_n')V_n^m\bigr] + i\mathcal{C}_1[y_n U_n^m] = N_n^m,$$
$$\mathcal{C}_1\bigl[(\tfrac{1}{R}j_n + kj_n')V_n^m\bigr] + \mathcal{C}_2[j_n U_n^m] = 0,$$
$$i\mathcal{C}_1\bigl[(\tfrac{1}{R}y_n + ky_n')V_n^m\bigr] + i\mathcal{C}_2[y_n U_n^m] = \operatorname{curl} N_n^m.$$

Multiplication of the first equation with $iy_n = iy_n(kR)$, the second one with $j_n$ and subtraction yields, using the Wronskian $j_n y_n' - j_n' y_n = \frac{1}{k^2 R^2}$,

$$\frac{i}{k^3 R^2}\mathcal{C}_2[V_n^m] = j_n N_n^m.$$

A similar computation for the third and forth equation yields

$$\frac{i}{kR^2}\mathcal{C}_2[U_n^m] = -[\tfrac{1}{R}j_n + kj_n']\operatorname{curl} N_n^m.$$

## 2. The far fiel operator

From the last two equations we conclude for a vector $p \in \mathbb{C}^3$:

$$\int_{S^2} \mathrm{curl}_y^2 \left[ p\, \Phi(x, R\hat{y}) \right] \cdot U_n^m(\hat{y}) \, \mathrm{d}s(\hat{y})$$
$$= ik \left[ \tfrac{1}{R} j_n(kR) + k j_n'(kR) \right] p \cdot \mathrm{curl}\, N_n^m(x),$$

$$\int_{S^2} \mathrm{curl}_y^2 \left[ p\, \Phi(x, R\hat{y}) \right] \cdot V_n^m(\hat{y}) \, \mathrm{d}s(\hat{y}) = -ik^3 j_n(kR) p \cdot N_n^m(x)$$

or

$$\int_{S^2} \mathrm{curl}_y^2 \left[ p\, \Phi(x, R\hat{y}) \right] \cdot U_n^m(\hat{y}) \, \mathrm{d}s(\hat{y})$$
$$= ik \left[ \tfrac{1}{R} j_n(kR) + k j_n'(kR) \right] \left[ \tfrac{1}{|x|} h_n(k|x|) + k h_n'(k|x|) \right] p \cdot U_n^m(\hat{x}) \quad (4.7)$$

and

$$\int_{S^2} \mathrm{curl}_y^2 \left[ p\, \Phi(x, R\hat{y}) \right] \cdot V_n^m(\hat{y}) \, \mathrm{d}s(\hat{y}) = ik^3 j_n(kR) h_n(k|x|) p \cdot V_n^m(\hat{x}) \tag{4.8}$$

The terms depending on $x$ and $|x|$, respectively, have the following asymptotic behaviour:

$$\mathrm{curl}_y^2 \left[ p\Phi(x, R\hat{y}) \right] = \frac{k^2}{4\pi |x|} e^{ik|x|} (\hat{x} \times p \times \hat{x}) e^{-ikR\hat{x}\cdot\hat{y}} + \mathcal{O}(|x|^{-2}),$$

$$h_n(k|x|) = \frac{1}{i^{n+1} k |x|} e^{ik|x|} + \mathcal{O}(|x|^{-2}),$$

$$h_n'(k|x|) = \frac{1}{i^n k |x|} e^{ik|x|} + \mathcal{O}(|x|^{-2}).$$

Hence, letting $|x| \to \infty$ in equations (4.7) and (4.8) yields

$$\int_{S^2} e^{-ikR\hat{x}\cdot\hat{y}} (\hat{x} \times p \times \hat{x}) \cdot U_n^m(\hat{y}) \, \mathrm{d}s(\hat{y})$$
$$= \frac{i}{k} \frac{4\pi}{i^n} \left[ \tfrac{1}{R} j_n(kR) + k j_n'(kR) \right] p \cdot U_n^m(\hat{x}),$$

$$\int_{S^2} e^{-ikR\hat{x}\cdot\hat{y}} (\hat{x} \times p \times \hat{x}) \cdot V_n^m(\hat{y}) \, \mathrm{d}s(\hat{y}) = \frac{4\pi}{i^n} j_n(kR)\, p \cdot V_n^m(\hat{x}).$$

Now we can go back to the expansion of a plane wave:
$$pe^{ik\,d\cdot x} = \sum a_n^m M_n^m(x,k) + b_n^m \tfrac{1}{ik}\operatorname{curl} M_n^m(x,k)$$
with
$$-a_n^m j_n(k|x|) = \int_{S^2} e^{ik|x|\,\hat{y}\cdot d}(d\times p\times d)\cdot \overline{V_n^m(\hat{y})}\,\mathrm{d}s(\hat{y}),$$
$$\tfrac{1}{ik}b_n^m\bigl[\tfrac{1}{|x|}j_n(k|x|) + kj_n'(k|x|)\bigr] = \int_{S^2} e^{ik|x|\,\hat{y}\cdot d}(d\times p\times d)\cdot \overline{U_n^m(\hat{y})}\,\mathrm{d}s(\hat{y}).$$

We compute
$$\overline{\int_{S^2} e^{ik|x|\,\hat{y}\cdot d}(d\times p\times d)\cdot \overline{V_n^m(\hat{y})}\,\mathrm{d}s(\hat{y})}$$
$$= \int_{S^2} e^{-ik|x|\,\hat{y}\cdot d}(d\times \overline{p}\times d)\cdot V_n^m(\hat{y})\,\mathrm{d}s(\hat{y})$$
$$= \frac{4\pi}{i^n} j_n(k|x|)\,\overline{p}\cdot V_n^m(d)$$

and
$$\overline{\int_{S^2} e^{ik|x|\,\hat{y}\cdot d}(d\times p\times d)\cdot \overline{U_n^m(\hat{y})}\,\mathrm{d}s(\hat{y})}$$
$$= \int_{S^2} e^{-ik|x|\,\hat{y}\cdot d}(d\times \overline{p}\times d)\cdot U_n^m(\hat{y})\,\mathrm{d}s(\hat{y})$$
$$= \frac{i}{k}\frac{4\pi}{i^n}\bigl[\tfrac{1}{|x|}j_n(k|x|) + kj_n'(k|x|)\bigr]\overline{p}\cdot U_n^m(d).$$

Hence
$$a_n^m = -4\pi i^n\, p\cdot \overline{V_n^m(d)},$$
$$b_n^m = 4\pi i^n\, p\cdot \overline{U_n^m(d)}.$$

Finally,
$$pe^{ikx\cdot d} = 4\pi \sum i^n \bigl[\tfrac{1}{ik} p\cdot \overline{U_n^m(d)}\operatorname{curl} M_n^m(x) - p\cdot \overline{V_n^m(d)} M_n^m(x)\bigr].$$

As corollary we find the series expansion of the plane wave of the form $(d\times p)e^{ik\,d\cdot x} = \tfrac{1}{ik}\operatorname{curl}[p\,e^{ik\,d\cdot x}]$, namely
$$(d\times p)e^{ikx\cdot d} = \sum a_n^m M_n^m(x,k) + b_n^m \tfrac{1}{ik}\operatorname{curl} M_n^m(x,k)$$

2. The far fiel operator                                                                99

with
$$a_n^m = -4\pi i^n \, p \cdot \overline{U_n^m(d)},$$
$$b_n^m = -4\pi i^n \, p \cdot \overline{V_n^m(d)}.$$

## 2.2. Achiral case

For the inverse problem we consider plane waves as incident fields: $H^i(x,d,p) = p\, e^{ik\,x\cdot d}$. We computed the series expansion of such a plane wave with direction of incidence $d$ and polarization $p$,

$$H^i(x;d,p) = 4\pi \sum \frac{i^n}{ik} p \cdot \overline{U_n^m(d)} \operatorname{curl} M_n^m(x,k) - i^n p \cdot \overline{V_n^m(d)} M_n^m(x,k).$$

The incident field is scattered by a sphere $B(0,1)$ with wave number $\kappa$ in the interior and $k$ in the exterior. The corresponding far field pattern $H^\infty(\hat{x};d,p)$ of the scattered field caused by $H^i$ is given by the series

$$\frac{(4\pi)^2 i}{k} \sum \frac{\operatorname{Re} \det_n(\kappa)}{\det_n(\kappa)} \left[ p \cdot \overline{V_n^m(d)} V_n^m(\hat{x}) + p \cdot \overline{U_n^m(d)} U_n^m(\hat{x}) \right].$$

Recall the definition of the far field operator:

$$\mathcal{F}\colon L^2_t(\mathbb{S}^2) \to L^2_t(\mathbb{S}^2), \quad p \mapsto \int_{\mathbb{S}^2} H^\infty\bigl(\hat{x}; \theta, p(\theta)\bigr) \,\mathrm{d}s(\theta).$$

Then
$$\mathcal{F}p = \frac{(4\pi)^2 i}{k} \sum \frac{\operatorname{Re} \det_n(\kappa)}{\det_n(\kappa)} \left[ p_n^m V_n^m + q_n^m U_n^m \right]$$

where the Fourier coefficients

$$p_n^m := \int_{\mathbb{S}^2} p(\theta) \cdot \overline{V_n^m(\theta)} \,\mathrm{d}s(\theta) \quad \text{and} \quad q_n^m := \int_{\mathbb{S}^2} p(\theta) \cdot \overline{U_n^m(\theta)} \,\mathrm{d}s(\theta)$$

and the tangential field $p \in L^2_t(\mathbb{S}^2)$ has the expansion

$$p = \sum p_n^m V_n^m + q_n^m U_n^m.$$

Finally, we can determine an eigensystem of $\mathcal{F}$. In the achiral case the eigenvalues of $\mathcal{F}$ are given by

$$\lambda_n = \frac{(4\pi)^2}{k} \frac{\operatorname{Re} \det_n(\kappa)}{\det_n(\kappa)}, \quad n = 0, 1, 2, \ldots$$

They have the multiplicity $2n+1$ and the vector spherical harmonics $U_n^m$ and $V_n^m$ are the eigenfunctions. Now, we can compute the series

$$\sum_{n\in\mathbb{N}}\sum_{m=-n}^{n}\frac{|(\phi_z,U_n^m)_{L_t^2(\mathbb{S}^2)}|^2}{|\lambda_n|}+\sum_{n\in\mathbb{N}}\sum_{m=-n}^{n}\frac{|(\phi_z,V_n^m)_{L_t^2(\mathbb{S}^2)}|^2}{|\lambda_n|} \qquad (4.9)$$

which appears in Theorem III.34. For $z\in\mathbb{R}^3$, we choose

$$\phi_z(\hat{x}):=-ik\big[(\hat{x}\times z\times\hat{x})+(\hat{x}\times z)\big]e^{-ik\hat{x}\cdot z}.$$

Then $\phi_z=\operatorname{Grad}_{\hat{x}}\big[e^{-ik\hat{x}\cdot z}\big]+\hat{x}\times\operatorname{Grad}_{\hat{x}}\big[e^{-ik\hat{x}\cdot z}\big]$ and we already know the series representation of $e^{-ik\hat{x}\cdot z}$ from the Jacobi–Anger expansion, namely

$$e^{-ik\hat{x}\cdot z}=4\pi\sum(-i)^n j_n(k|z|)\overline{Y_n^m(\hat{z})}Y_n^m(\hat{x}).$$

Hence, $\phi_z$ has the series expansion

$$\phi_z(\hat{x})=4\pi\sum(-i)^n j_n(k|z|)\overline{Y_n^m(\hat{z})}\big[\operatorname{Grad}Y_n^m(\hat{x})+\hat{x}\times\operatorname{Grad}Y_n^m(\hat{x})\big].$$

Recall the definition of $U_n^m(\hat{x})=1/\sqrt{n(n+1)}\operatorname{Grad}Y_n^m(\hat{x})$ and $V_n^m(\hat{x})=\hat{x}\times U_n^m(\hat{x})$. The Fourier coefficients of $\phi_z$ are given by

$$(\phi_z,U_n^m)_{L_t^2(\mathbb{S}^2)}=(\phi_z,V_n^m)_{L_t^2(\mathbb{S}^2)}=4\pi(-i)^n j_n(k|z|)\overline{Y_n^m(\hat{z})}.$$

As in the scalar case (compare section 1.5 in Kirsch [24])

$$\sum_{m=-n}^{n}|(\phi_z,U_n^m)_{L_t^2(\mathbb{S}^2)}|^2=4\pi(2n+1)\frac{(k|z|)^{2n}}{[(2n+1)!!]^2}(1+\mathcal{O}(1/n))$$

and

$$\sum_{m=-n}^{n}|(\phi_z,V_n^m)_{L_t^2(\mathbb{S}^2)}|^2=4\pi(2n+1)\frac{(k|z|)^{2n}}{[(2n+1)!!]^2}(1+\mathcal{O}(1/n)).$$

Here $p!!:=1\cdot 3\cdot 5\cdots p$ for any odd number $p$. We continue with the asymptotic behavior of the eigenvalues

$$\lambda_n=\frac{(4\pi)^2 i}{k}\frac{\operatorname{Re}\det_n(\kappa)}{\det_n(\kappa)}=\frac{(4\pi)^2 i}{k}\cdot\frac{j_n(k)}{h_n(k)}\cdot\frac{\left(\frac{1}{\kappa}-\frac{1}{k}\right)+\frac{j_n'(\kappa)}{j_n(\kappa)}-\frac{j_n'(k)}{j_n(k)}}{\left(\frac{1}{\kappa}-\frac{1}{k}\right)+\frac{j_n'(\kappa)}{j_n(\kappa)}-\frac{h_n'(k)}{h_n(k)}}.$$

## 2. The far fiel operator

$j_n, h_n, j'_n$ and $h'_n$ have the following asymptotic behavior:

$$j_n(t) = \frac{t^n}{(2n+1)!!}(1+\mathcal{O}(\tfrac{1}{n})), \quad j'_n(t) = \frac{nt^{n-1}}{(2n+1)!!}(1+\mathcal{O}(\tfrac{1}{n})),$$

$$h_n(t) = \frac{(2n-1)!!}{it^{n+1}}(1+\mathcal{O}(\tfrac{1}{n})), \quad h'_n(t) = -\frac{(n+1)(2n-1)!!}{it^{n+2}}(1+\mathcal{O}(\tfrac{1}{n})).$$

Plugging this into $1/\lambda_n$ yields

$$\frac{1}{\lambda_n} = \frac{(2n+1)!!(2n-1)!!}{(4\pi)^2 k^{2n}} \cdot \frac{\left(\frac{1}{\kappa}-\frac{1}{k}\right)+\frac{n}{\kappa}+\frac{n+1}{k}}{\left(\frac{1}{\kappa}-\frac{1}{k}\right)+\frac{n}{\kappa}-\frac{n}{k}}(1+\mathcal{O}(\tfrac{1}{n})).$$

The second fraction on the right can be simplified and we get

$$\frac{\left(\frac{1}{\kappa}-\frac{1}{k}\right)+\frac{n}{\kappa}+\frac{n+1}{k}}{\left(\frac{1}{\kappa}-\frac{1}{k}\right)+\frac{n}{\kappa}-\frac{n}{k}} = 1 - \frac{\kappa}{(n+1)(k-\kappa)}.$$

Hence,

$$\frac{1}{\lambda_n} = \frac{(2n+1)!!(2n-1)!!}{(4\pi)^2 k^{2n}}(1+\mathcal{O}(1/n))$$

and

$$\sum_{m=-n}^{n} \frac{|(\phi_z, U_n^m)_{L_t^2(\mathbb{S}^2)}|^2}{|\lambda_n|} = \sum_{m=-n}^{n} \frac{|(\phi_z, V_n^m)_{L_t^2(\mathbb{S}^2)}|^2}{|\lambda_n|} = \frac{|z|^{2n}}{4\pi}(1+\mathcal{O}(1/n)).$$

We conclude that the series (4.9) converges if, and only if, $|z|<1$; that is, $z$ is inside the ball $B(0,1)$.

### 2.3. Chiral case

In the chiral case we find an explicit form for $\mathcal{F}$ in a similar way: Given the tangential field $p \in L_t^2(\mathbb{S}^2)$: $p = \sum p_n^m U_n^m + q_n^m V_n^m$ with

$$p_n^m = \int_{\mathbb{S}^2} p(\theta) \cdot \overline{U_n^m(\theta)}\,\mathrm{d}s(\theta) \quad \text{and} \quad q_n^m = \int_{\mathbb{S}^2} p(\theta) \cdot \overline{V_n^m(\theta)}\,\mathrm{d}s(\theta)$$

we note that

$$p = \sum p_n^m U_n^m + q_n^m V_n^m$$
$$= \frac{1}{2}\sum (p_n^m - iq_n^m)(U_n^m + iV_n^m) + (p_n^m + iq_n^m)(U_n^m - iV_n^m).$$

In the chiral case two constants $\kappa_L = \frac{\kappa}{1-\kappa\beta}$ and $\kappa_R = \frac{\kappa}{1+\kappa\beta}$ appeared as some kind of wave number for the fields $Q_L$ and $Q_R$. We computed the far field pattern to be

$$H^\infty(\hat{x}) = \frac{4\pi}{2ik} \sum \frac{c_L}{i^{n+1}} \left[ (\beta_n^m + i\alpha_n^m) V_n^m(\hat{x}) + (\alpha_n^m - i\beta_n^m) U_n^m(\hat{x}) \right]$$
$$- \frac{c_R}{i^{n+1}} \left[ (\beta_n^m - i\alpha_n^m) V_n^m(\hat{x}) + (\alpha_n^m + i\beta_n^m) U_n^m(\hat{x}) \right]$$

with coefficients $\alpha_n^m = -4\pi i^n p \cdot \overline{V_n^m(d)}$, $\beta_n^m = 4\pi i^n p \cdot \overline{U_n^m(d)}$ and the constants $c_L = \frac{\text{Re det}_n(\kappa_L)}{\det_n(\kappa_L)}$, $c_R = \frac{\text{Re det}_n(\kappa_R)}{\det_n(\kappa_R)}$. As in the achiral case, we conclude

$$\mathcal{F}p = \frac{(4\pi)^2 i}{2k} \sum c_L(p_n^m - iq_n^m)(U_n^m + iV_n^m) - c_R(p_n^m + iq_n^m)(U_n^m - iV_n^m).$$

We observe that $U_n^m + iV_n^m$, $m = -n, \ldots, n$, are eigenfunctions for the eigenvalue

$$\lambda_n = \frac{(4\pi)^2 i}{k} \frac{\text{Re det}_n(\kappa_L)}{\det_n(\kappa_L)}$$

and $U_n^m - iV_n^m$, $m = -n, \ldots, n$, are eigenfunctions for the eigenvalue

$$\lambda_n = -\frac{(4\pi)^2 i}{k} \frac{\text{Re det}_n(\kappa_R)}{\det_n(\kappa_R)}$$

for $n \in \mathbb{N}_0$. Again the eigenvalues have multiplicity $2n+1$. The evaluation of the series

$$\sum_{j \in \mathbb{N}} \frac{|(\phi_z, \psi_j)_{L_t^2(\mathbb{S}^2)}|^2}{|\lambda_j|}$$

for the characteristic function of the scatterer is completely analogous to the achiral case. We skip the computation at this point.

CHAPTER V

# Factorization Method for the vector Helmholtz equation

This chapter can be seen as an application of the results from chapters II and III. The equations and operators look different but the concepts and the arguments of the proofs are the same. That is why we present this chapter not as detailed as the other two. Nevertheless, we treat the subject rigorously.

Scattering by an infinite chiral cylinder leads to a scattering problem for the vector Helmholtz equation, comparable with problems that lead to the scalar Helmholtz equation for TE– or TM–modes. The first section describes the setting and deduces the vector Helmholtz equation which is studied in sections 2 and 3. Here, we apply the methods already used for scattering by a bounded obstacle: We formulate a variational equation, deduce an equivalent integro–differential equation and show existence und uniqueness. The Factorization method is adapted for the inverse problem. In section 4 we study conditions for the material parameters $\varepsilon$, $\mu$ and $\beta$ such that the rather abstract assumptions for solvability and the Factorization method are fullfilled. Finally, we present numerical experiments concerning plots of the far field pattern and the reconstruction of the cylinder.

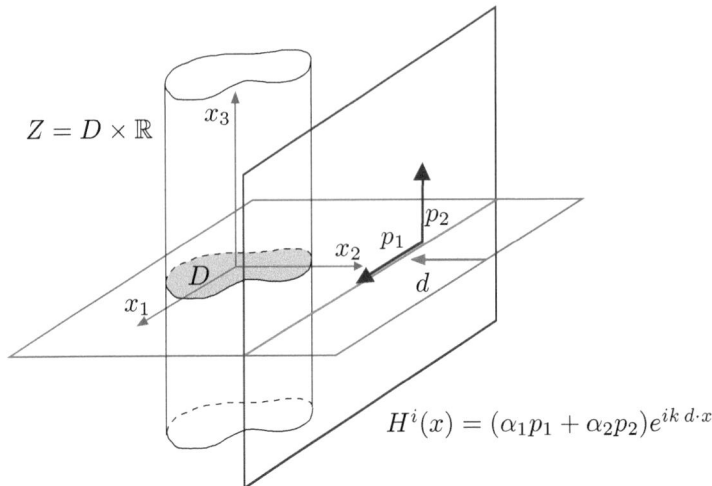

Figure V.1: Direct problem setting.

## 1. Motivation: Scattering by a chiral cylinder

In this chapter we study the scattering by an infinite chiral cylinder $Z$ along the $x_3$-axis. We choose incident fields which are orthogonal to the cylinder; that is, the vector $d$ – direction of incidence – lies in the $(x_1, x_2)$-plane. Given a bounded domain $D \subset \mathbb{R}^2$ then $Z$ is simply $D \times \mathbb{R}$. Figure V.1 shows the cylinder and the vectors characterizing the incident field $H^i$ (and $E^i$): the direction of incidence $d$ and two linear independent polarization vectors $p_1, p_2$ in the plane which is orthogonal to $d$. The polarization of $H^i$ is a linear combination of the basis vectors $p_1, p_2$.

The cylinder $Z$ is characterized by the electric permittivity $\varepsilon$, the magnetic permeability $\mu$ and the chirality $\beta$ in the following way. We assume that $\varepsilon, \mu$ and $\beta$ do not depend on $x_3$. $\varepsilon, \mu, \beta \colon \mathbb{R}^2 \to \mathbb{C}$ with $\varepsilon \equiv \varepsilon_0$, $\mu \equiv \mu_0$ and $\beta \equiv 0$ in $\mathbb{R}^2 \setminus \overline{D}$.

# 1. Motivation: Scattering by a chiral cylinder

## Total field

We start with Maxwell's equations (2.5), (2.6); that is,

$$\operatorname{curl} H = -ik\varepsilon\big(E + \beta \operatorname{curl} E\big), \tag{5.1}$$
$$\operatorname{curl} E = ik\mu\big(H + \beta \operatorname{curl} H\big) \tag{5.2}$$

with the wave number $k^2 = \omega^2 \varepsilon_0 \mu_0$ and the relative paramters $\varepsilon$ and $\mu$. Since the direction of incidence is orthogonal to $Z$ the incident field does not depend on $x_3$ either. Hence, the whole system is invariant along $x_3$ and the $x_3$-derivatives vanish. This yields:

$$\frac{\partial H_3}{\partial x_2} = -ik\varepsilon \left( E_1 + \beta \frac{\partial E_3}{\partial x_2} \right), \tag{5.3}$$

$$-\frac{\partial H_3}{\partial x_1} = -ik\varepsilon \left( E_2 - \beta \frac{\partial E_3}{\partial x_1} \right), \tag{5.4}$$

$$\frac{\partial H_2}{\partial x_1} - \frac{\partial H_1}{\partial x_2} = -ik\varepsilon \left( E_3 + \beta \left( \frac{\partial E_2}{\partial x_1} - \frac{\partial E_1}{\partial x_2} \right) \right) \tag{5.5}$$

and

$$\frac{\partial E_3}{\partial x_2} = ik\mu \left( H_1 + \beta \frac{\partial H_3}{\partial x_2} \right), \tag{5.6}$$

$$-\frac{\partial E_3}{\partial x_1} = ik\mu \left( H_2 - \beta \frac{\partial H_3}{\partial x_1} \right), \tag{5.7}$$

$$\frac{\partial E_2}{\partial x_1} - \frac{\partial E_1}{\partial x_2} = ik\mu \left( H_3 + \beta \left( \frac{\partial H_2}{\partial x_1} - \frac{\partial H_1}{\partial x_2} \right) \right). \tag{5.8}$$

We can reduce this system of six equations to two coupled equations for the components $E_3$ and $H_3$: We differentiate (5.3) with respect to $x_2$ and (5.4) with respect to $x_1$. Subtraction yields

$$\frac{\partial E_2}{\partial x_1} - \frac{\partial E_1}{\partial x_2} = \frac{\partial}{\partial x_1}\left(\beta \frac{\partial E_3}{\partial x_1}\right) + \frac{\partial}{\partial x_2}\left(\beta \frac{\partial E_3}{\partial x_2}\right)$$
$$- \frac{\partial}{\partial x_1}\left(\frac{i}{k\varepsilon} \frac{\partial H_3}{\partial x_1}\right) - \frac{\partial}{\partial x_2}\left(\frac{i}{k\varepsilon} \frac{\partial H_3}{\partial x_2}\right)$$
$$= \operatorname{div}\left(\beta \nabla E_3\right) - \frac{i}{k}\operatorname{div}\left(\frac{1}{\varepsilon}\nabla H_3\right).$$

… Factorization Method for the vector Helmholtz case

Analogously, we deduce from (5.6) and (5.7)

$$\frac{\partial H_2}{\partial x_1} - \frac{\partial H_1}{\partial x_2} = \frac{\partial}{\partial x_1}\left(\beta\frac{\partial H_3}{\partial x_1}\right) + \frac{\partial}{\partial x_2}\left(\beta\frac{\partial H_3}{\partial x_2}\right)$$
$$+ \frac{\partial}{\partial x_1}\left(\frac{i}{k\mu}\frac{\partial E_3}{\partial x_1}\right) + \frac{\partial}{\partial x_2}\left(\frac{i}{k\mu}\frac{\partial E_3}{\partial x_2}\right)$$
$$= \operatorname{div}\left(\beta\,\nabla H_3\right) + \frac{i}{k}\operatorname{div}\left(\frac{1}{\mu}\nabla E_3\right).$$

Here the differential operators are $\nabla = (\frac{\partial}{\partial x_1}, \frac{\partial}{\partial x_2})^\top$ and $\operatorname{div} v = \nabla \cdot v$. (The derivatives with respect to $x_3$ vanish anyway.) Plugging these two expressions into (5.5) and (5.8) yields

$$\operatorname{div}\left(\beta\,\nabla H_3\right) + \tfrac{i}{k}\operatorname{div}\left(\tfrac{1}{\mu}\nabla E_3\right)$$
$$= -ik\varepsilon E_3 - ik\varepsilon\beta\left[\operatorname{div}\left(\beta\,\nabla E_3\right) - \tfrac{i}{k}\operatorname{div}\left(\tfrac{1}{\varepsilon}\nabla H_3\right)\right] \quad (5.9)$$

and

$$\operatorname{div}\left(\beta\,\nabla E_3\right) - \tfrac{i}{k}\operatorname{div}\left(\tfrac{1}{\varepsilon}\nabla H_3\right)$$
$$= ik\mu H_3 + ik\mu\beta\left[\operatorname{div}\left(\beta\,\nabla H_3\right) + \tfrac{i}{k}\operatorname{div}\left(\tfrac{1}{\mu}\nabla E_3\right)\right]. \quad (5.10)$$

We multiply equation (5.10) by $ik\varepsilon\beta$ and subtract equation (5.9) from it:

$$(1 - k^2\varepsilon\mu\beta^2)\left[\operatorname{div}\left(\beta\,\nabla H_3\right) + \tfrac{i}{k}\operatorname{div}\left(\tfrac{1}{\mu}\nabla E_3\right)\right] + ik\varepsilon E_3 - k^2\varepsilon\mu\beta H_3 = 0.$$

Analogously we multiply (5.9) by $ik\mu\beta$ and add it to (5.10):

$$(1 - k^2\varepsilon\mu\beta^2)\left[\operatorname{div}\left(\beta\,\nabla E_3\right) - \tfrac{i}{k}\operatorname{div}\left(\tfrac{1}{\varepsilon}\nabla H_3\right)\right] - ik\mu H_3 - k^2\varepsilon\mu\beta E_3 = 0.$$

After multiplication with $-ik$ and $ik$, respectively, the last two equations can be written as system:

$$\operatorname{div}\left[\begin{pmatrix}\frac{1}{\mu} & -ik\beta \\ ik\beta & \frac{1}{\varepsilon}\end{pmatrix}\begin{pmatrix}\nabla E_3 \\ \nabla H_3\end{pmatrix}\right]$$
$$+ k^2\frac{1}{1-k^2\varepsilon\mu\beta^2}\begin{pmatrix}\varepsilon & ik\varepsilon\mu\beta \\ -ik\varepsilon\mu\beta & \mu\end{pmatrix}\begin{pmatrix}E_3 \\ H_3\end{pmatrix} = 0. \quad (5.11)$$

# 1. Motivation: Scattering by a chiral cylinder

Here div is applied to each component. Note that – more precisely – the first matrix is the $(4 \times 4)$–matrix

$$\begin{pmatrix} \frac{1}{\mu} & 0 & -ik\beta & 0 \\ 0 & \frac{1}{\mu} & 0 & -ik\beta \\ \hline ik\beta & 0 & \frac{1}{\varepsilon} & 0 \\ 0 & ik\beta & 0 & \frac{1}{\varepsilon} \end{pmatrix}.$$

We will use the following notation: Given a $(4 \times 4)$–matrix $A$ of the form

$$A = \begin{pmatrix} a_{11} & 0 & a_{12} & 0 \\ 0 & a_{11} & 0 & a_{12} \\ \hline a_{21} & 0 & a_{22} & 0 \\ 0 & a_{21} & 0 & a_{22} \end{pmatrix}, \qquad (5.12)$$

$\tilde{A}$ denotes the corresponding $(2 \times 2)$–matrix

$$\tilde{A} = \begin{pmatrix} a_{11} & a_{12} \\ a_{21} & a_{22} \end{pmatrix}.$$

Introducing $v := (E_3, H_3)^\top$ the system (5.11) has the form

$$\operatorname{div}(A\nabla v) + k^2 B v = 0$$

where $A \colon \mathbb{R}^2 \to \mathbb{C}^{4\times 4}$ of the form (5.12) and $B \colon \mathbb{R}^2 \to \mathbb{C}^{2\times 2}$ are matrix functions with $A \equiv I_4$ and $B \equiv I_2$ in $\mathbb{R}^2 \setminus \overline{D}$. $I = I_n$ denotes the $n \times n$–unit matrix. (We will skip the index.) $v \colon \mathbb{R}^2 \to \mathbb{C}^2$ and $\nabla v := (\nabla v_1, \nabla v_2)^\top$ and for $w = (w_1, w_2)^\top$ with two functions $w_1, w_2 \colon \mathbb{R}^2 \to \mathbb{C}^2$ define $\operatorname{div} w := (\operatorname{div} w_1, \operatorname{div} w_2)^\top$.

**Incident field**

As incident fields we choose plane waves orthogonal to $Z$. Hence, with polarization vector $p$ and direction of incidence $d$ with $p \cdot d = 0$ (divergence free fields)

$$H^i(x) = p e^{ikd \cdot x} \qquad \text{and} \qquad E^i(x) = -(d \times p)\, e^{ikd \cdot x} \qquad (5.13)$$

where $d = (d_1, d_2, 0)^\top$ and $|d| = 1$. We only need the third components $H_3^i$ and $E_3^i$ and show that they are independent of each other: Given $r, s \in \mathbb{C}$

let $H_3^i(x) = re^{ikd\cdot x}$ and $E_3^i(x) = -se^{ikd\cdot x}$. Then for $p \in \mathbb{C}^3$ defined by $p := (-d_2 s, d_1 s, r)^\top$ we have $p\cdot d = 0$ and the representation (5.13) holds. Furthermore $v^i := (E_3^i, H_3^i)^\top$ solves the vector Helmholtz equation

$$\Delta v^i + k^2 v^i = 0 \quad \text{in } \mathbb{R}^2.$$

**Scattered field**

The total field $v$ is the sum of the incident and the scattered field,

$$v = v^s + v^i$$

where $v^i$ is a solution of the vector Helmholtz equation. Hence, the scattered field $v^s$ solves

$$\operatorname{div}(A\nabla v^s) + k^2 B v^s = \operatorname{div}\left[(I-A)\nabla v^i\right] - k^2(B-I)v^i \qquad (5.14)$$

in $\mathbb{R}^2$. Note that the right–hand side of this equations can be interpreted as source $\operatorname{div} f - k^2 g$ with functions $f, g$ whose support is contained in $\overline{D}$.

**Transmission conditions**

Now we deduce the transmission conditions. The third component of the unit normal vector of $Z$ vanishes: $\nu = (\nu_1, \nu_2, 0)^\top$. From $\nu \times E_+ = \nu \times E_-$ on interfaces we deduce

$$\begin{pmatrix} \nu_2 E_{3+} \\ -\nu_1 E_{3+} \\ \nu_1 E_{2+} - \nu_2 E_{1+} \end{pmatrix} = \begin{pmatrix} \nu_2 E_{3-} \\ -\nu_1 E_{3-} \\ \nu_1 E_{2-} - \nu_2 E_{1-} \end{pmatrix}.$$

Since $|\nu| = 1$ the first two equations yield $E_{3+} = E_{3-}$ and analogously $H_{3+} = H_{3-}$ or – in terms of $v^s$ –

$$v_+^s = v_-^s \quad \text{on } \partial D. \qquad (5.15)$$

In order to deduce the transmission condition for the normal derivatives we easily compute the third component of $\nu \times \operatorname{curl} E$:

$$(\nu \times \operatorname{curl} E)_3 = -\frac{\partial E_3}{\partial x_1}\nu_1 - \frac{\partial E_3}{\partial x_2}\nu_2 = -\frac{\partial E_3}{\partial \nu}.$$

We rewrite equations (5.1), (5.2)

$$ikH = \tfrac{1}{\mu}\operatorname{curl} E - ik\beta \operatorname{curl} H,$$
$$-ikE = \tfrac{1}{\varepsilon}\operatorname{curl} H + ik\beta \operatorname{curl} E$$

and using $\nu \times E_+ = \nu \times E_-$ and $\nu \times H_+ = \nu \times H_-$ again yields

$$\nu \times \operatorname{curl} E_+ = \tfrac{1}{\mu_-}\nu \times \operatorname{curl} E_- - ik\beta_- \nu \times \operatorname{curl} H_-,$$
$$\nu \times \operatorname{curl} H_+ = \tfrac{1}{\varepsilon_-}\nu \times \operatorname{curl} H_- + ik\beta_- \nu \times \operatorname{curl} E_-$$

on $\partial D$. We take only the third component of each equation and get the transmission condition

$$\begin{pmatrix} \frac{\partial E_3}{\partial \nu}{}_+ \\ \frac{\partial H_3}{\partial \nu}{}_+ \end{pmatrix} = \begin{pmatrix} \frac{1}{\mu_-} & -ik\beta_- \\ ik\beta_- & \frac{1}{\varepsilon_-} \end{pmatrix} \begin{pmatrix} \frac{\partial E_3}{\partial \nu}{}_- \\ \frac{\partial H_3}{\partial \nu}{}_- \end{pmatrix} \quad \text{on } \partial D.$$

In terms of $v^s$ with $\frac{\partial v}{\partial \nu} = (\frac{\partial v_1}{\partial \nu}, \frac{\partial v_2}{\partial \nu})^\top$ this condition reads

$$\tilde{A}_- \frac{\partial v^s}{\partial \nu}{}_- - \frac{\partial v^s}{\partial \nu}{}_+ = (I - \tilde{A}_-)\frac{\partial v^i}{\partial \nu} \quad \text{on } \partial D. \tag{5.16}$$

**Radiation condition**

Finally, since we are interested in outgoing waves each component of $v^s$ shall satisfy the Sommerfeld radiation condition in $\mathbb{R}^2$; that is,

$$\frac{\partial v_j^s}{\partial r} - ikv_j^s = \mathcal{O}(r^{-3/2}), \qquad \text{for } r = |x| \to \infty, j = 1, 2$$

uniformly with respect to $\hat{x} = x/|x| \in \mathbb{S}^1$. $\mathbb{S}^1$ denotes the unit circle. Functions which satisfy this radiation condition will again be called RADIATING.

## 2. Direct transmission problem

We state the direct problem variationally and start with the divergence theorem which fits to our case. Let $D \subset \mathbb{R}^2$ be bounded with boundary of class $C^2$. Given sufficiently smooth functions $v_1, v_2, \psi_1, \psi_2 : \overline{D} \to \mathbb{C}$

and $A: \overline{D} \to \mathbb{C}^{4\times 4}$ of the form (5.12) we have the following form of the divergence theorem for $v = (v_1, v_2)^\top$, $\psi = (\psi_1, \psi_2)^\top$:

$$\iint_D \operatorname{div}(A\nabla v) \cdot \psi \, \mathrm{d}x = - \iint_D (A\nabla v) \cdot \nabla \psi \, \mathrm{d}x + \int_{\partial D} \left( \tilde{A} \frac{\partial v}{\partial \nu} \right) \cdot \psi \, \mathrm{d}s$$

where $\nabla v = (\nabla v_1, \nabla v_2)^\top$, the analog for $\nabla \psi$, $\frac{\partial v}{\partial \nu} = (\frac{\partial v_1}{\partial \nu}, \frac{\partial v_2}{\partial \nu})^\top$ and div is applied to each component. Indeed, with $\tilde{A} = (a_{jl})_{j,l=1,2}$

$$\iint_D \operatorname{div}(a_{11}\nabla v_1 + a_{12}\nabla v_2)\psi_1 + \operatorname{div}(a_{21}\nabla v_1 + a_{22}\nabla v_2)\psi_2 \, \mathrm{d}x$$

$$= - \iint_D (a_{11}\nabla v_1 + a_{12}\nabla v_2) \cdot \nabla \psi_1 + (a_{21}\nabla v_1 + a_{22}\nabla v_2) \cdot \nabla \psi_2 \, \mathrm{d}x$$

$$+ \int_{\partial D} (a_{11}\tfrac{\partial v_1}{\partial \nu} + a_{12}\tfrac{\partial v_2}{\partial \nu})\psi_1 + (a_{21}\tfrac{\partial v_1}{\partial \nu} + a_{22}\tfrac{\partial v_2}{\partial \nu})\psi_2 \, \mathrm{d}s.$$

by the (scalar) divergence theorem.

**Variational formulation**

We derive a variational formulation of the transmission problem found in the introductory part of this chapter. First recall the problem: Given a bounded domain $D \subset \mathbb{R}^2$ and an incident field $v^i = (v_1^i, v_2^i)^\top$ determine a solution $v^s = (v_1^s, v_2^s)^\top$ of the scattering equation (5.14) which satisfies the transmission conditions (5.15) and (5.16) on the boundary $\partial D$ and the Sommerfeld radiation condition; that is,

$$\operatorname{div}(A\nabla v^s) + k^2 B v^s = \operatorname{div}\left[(I-A)\nabla v^i\right] - k^2(B-I)v^i \quad \text{in } \mathbb{R}^2,$$

$$v_+^s = v_-^s \quad \text{on } \partial D,$$

$$\tilde{A}_- \frac{\partial v^s}{\partial \nu}_- - \frac{\partial v^s}{\partial \nu}_+ = (I - \tilde{A}_-)\frac{\partial v^i}{\partial \nu} \quad \text{on } \partial D$$

and

$$\frac{\partial v_j^s}{\partial r} - ikv_j^s = \mathcal{O}(r^{-3/2}) \quad \text{for } r = |x| \to \infty, \ j = 1, 2$$

uniformly with respect to $\hat{x} = x/|x| \in \mathbb{S}^1$. The matrix functions $A, B$ are such that $A \equiv I$ and $B \equiv I$ in $\mathbb{R}^2 \setminus \overline{D}$.

After scalar multiplication of the partial differential equation with a test function $\psi = (\psi_1, \psi_2)^\top$ with compact support we integrate over $\mathbb{R}^2$.

## 2. Direct transmission problem

By the above divergence theorem – taking into account the transmission conditions – this yields the variational equation

$$\iint_{\mathbb{R}^2} (A\nabla v^s) \cdot \nabla \psi - k^2 (Bv^s) \cdot \psi \, dx$$
$$= \iint_D k^2 ((B-I)v^i) \cdot \psi + ((I-A)\nabla v^i) \cdot \nabla \psi \, dx$$

for all $\psi$ with compact support. As in the second chapter we generalize the transmission problem and allow square integrable source functions $g \in L^2(D, \mathbb{C}^2), h \in L^2(D, \mathbb{C}^4)$. First we introduce the function spaces we will use.

**Definition V.1.** Let $D \subset \mathbb{R}^2$ a domain.

(a) $H^1(D, \mathbb{C}^2) := \{v = (v_1, v_2)^\top : D \to \mathbb{C}^2 | v_j \in H^1(D), j = 1, 2\}$
Define $\nabla v := (\nabla v_1, \nabla v_2)^\top$.

(b) $H^1_{\text{loc}}(\mathbb{R}^2, \mathbb{C}^2) := \{v \colon \mathbb{R}^2 \to \mathbb{C}^2 | \forall \text{ balls } B \subset \mathbb{R}^2 : v|_B \in H^1(B, \mathbb{C}^2)\}$

(c) The test function space

$$\{\psi \colon \mathbb{R}^2 \to \mathbb{C}^2 | \exists \text{ ball } B \subset \mathbb{R}^2 : \operatorname{supp} \psi \subset B, \psi|_B \in H^1(B, \mathbb{C}^2)\}$$

is denoted by $H^1_c(\mathbb{R}^2, \mathbb{C}^2)$.

(d) For $w = (w_1, w_2)^\top$ with $w_1, w_2 \in H^1(D, \mathbb{C}^2)$ define

$$\operatorname{div} w := (\operatorname{div} w_1, \operatorname{div} w_2)^\top.$$

**Assumption V.2 (Material parameters).** Let $D \subset \mathbb{R}^2$ be a bounded Lipschitz domain. We assume that the complex matrix functions $A$ and $B$ – $A \in L^\infty(\mathbb{R}^2, \mathbb{C}^{4\times 4})$, $B \in L^\infty(\mathbb{R}^2, \mathbb{C}^{2\times 2})$ – are such that $A$ has the form (5.12), $A \equiv I$ and $B \equiv I$ in $\mathbb{R}^2 \setminus \overline{D}$.

**Problem 5 (Weak transmission problem for vector Helmholtz equation).** Let $k \in \Pi$ and Assumption V.2 be satisfied. Given the source terms $g \in L^2(D, \mathbb{C}^2)$ and $h \in L^2(D, \mathbb{C}^4)$ determine $v \in H^1_{\text{loc}}(\mathbb{R}^2, \mathbb{C}^2)$ such that $v$ is radiating and satisfies

$$\iint_{\mathbb{R}^2} (A\nabla v) \cdot \nabla \psi - k^2(Bv) \cdot \psi \, dx = \iint_D k^2 g \cdot \psi + h \cdot \nabla \psi \, dx \quad (5.17)$$

for all $\psi \in H^1_c(\mathbb{R}^2, \mathbb{C}^2)$.

Recall that $\Pi = \{z \in \mathbb{C} : z \neq 0, \operatorname{Re}(z) \geq 0, \operatorname{Im}(z) \geq 0\}$. To study solvability we deduce an equivalent integro–differential equation. In this case we need the fundamental solution to the scalar Helmholtz equation in $\mathbb{R}^2$. See section 3.10 in Colton and Kress [14]. We use the symbol $\Phi_\kappa$ again.

**Definition V.3 (Fundamental solution).** For $\kappa \in \Pi$ the fundamental solution $\Phi_\kappa$ to the scalar Helmholtz equation in $\mathbb{R}^2$

$$\Delta u + \kappa^2 u = 0$$

is defined by

$$\Phi_\kappa(x,y) := \tfrac{i}{4} H_0^{(1)}(\kappa|x-y|) \quad \text{for } x \neq y$$

where $H_0^{(1)}$ denotes the Hankel function of the first kind of order zero.

**Remark V.4.** It is well known that $\Phi_\kappa$ has a singularity at $x = y$ of the form $\log(\kappa|x-y|)$ (see Abramovitz and Stegun [1]). Furthermore, by the asymptotic behavior of $H_0^{(1)}$ the fundamental solution $\Phi_k$ satisfies the (two dimensional) Sommerfeld radiation condition.

We proceed with a lemma which provides the vector potentials leading to our IDE.

**Lemma V.5.** Let $\kappa \in \Pi$.

*(a)* For $g \in L^2(D, \mathbb{C}^2)$ the vector field

$$u(x) = \iint_D g(y) \Phi_\kappa(x,y) \, \mathrm{d}y, \quad x \in \mathbb{R}^2,$$

defines a function in $H^1_{\mathrm{loc}}(\mathbb{R}^2, \mathbb{C}^2)$ which is a weak solution of the vector Helmholtz equation $\Delta u + \kappa^2 u = -g$; that is,

$$\iint_{\mathbb{R}^2} \nabla u \cdot \nabla \psi - \kappa^2 u \cdot \psi \, \mathrm{d}x = \iint_D g \cdot \psi \, \mathrm{d}x$$

for all $\psi \in H^1_c(\mathbb{R}^2, \mathbb{C}^2)$. Furthermore $u$ is radiating and the restriction $u|_D$ of $u$ to $D$ defines a bounded operator from $L^2(D, \mathbb{C}^2)$ into $H^2(D, \mathbb{C}^2) := H^2(D) \times H^2(D)$.

## 2. Direct transmission problem

(b) For $h_1, h_2 \in L^2(D, \mathbb{C}^2)$ the vector field $u = (u_1, u_2)^\top$ with

$$u_j(x) = -\operatorname{div} \iint_D h_j(y) \Phi_\kappa(x, y) \, dy, \qquad j = 1, 2,$$

for $x \in \mathbb{R}^2$, defines a function in $H^1_{\text{loc}}(\mathbb{R}^2, \mathbb{C}^2)$ which is a weak solution of the Helmholtz equation $\Delta u + \kappa^2 u = \operatorname{div} h$; that is,

$$\iint_{\mathbb{R}^2} \nabla u_j \cdot \nabla \psi - \kappa^2 u_j \cdot \psi \, dx = \iint_D h_j \cdot \nabla \psi \, dx, \qquad j = 1, 2,$$

for all $\psi \in H^1_c(\mathbb{R}^2, \mathbb{C}^2)$. Furthermore $u$ is radiating and the restriction $u|_D$ of $u$ to $D$ defines a bounded operator from $L^2(D, \mathbb{C}^4)$ into $H^1(D, \mathbb{C}^2)$.

*Proof.* We use part (a) from Lemma 2.2 in Kirsch [21]. More precisely we need the two dimensional version whose proof will be absolutely analogous. Each component represents a two dimensional scalar Riesz potential, whence part (a).

Part (b) is just the vector version of the scalar case discussed in Lemmata 2.1 and 2.2 in Kirsch [23]. Here again we use a two dimensional version. □

Now we can reformulate the scattering equation (5.17) of the weak transmission problem

$$\iint_{\mathbb{R}^2} \nabla v \cdot \nabla \psi - k^2 v \cdot \psi \, dx = \iint_D k^2 [Q v + g] \cdot \psi + [P \nabla v + h] \cdot \nabla \psi \, dx$$

where the contrasts $P := I - A$ and $Q := B - I$. Using our vector potentials from the above lemma $v$ satisfies the following integro–differential equation

$$\begin{aligned} v(x) = k^2 &\iint_D [Q(y) v(y) + g(y)] \Phi_k(x, y) \, dy \\ -\operatorname{div} &\iint_D [P(y) \nabla v(y) + h(y)] \Phi_k(x, y) \, dy \end{aligned} \tag{5.18}$$

for $x \in D$.

**Theorem V.6 (Equivalence).** *(a) Let $v \in H^1_{\text{loc}}(\mathbb{R}^2, \mathbb{C}^2)$ be a radiating solution of (5.17). Then $v|_D \in H^1(D, \mathbb{C}^2)$ solves (5.18).*

*(b) Let $v \in H^1(D, \mathbb{C}^2)$ be a solution of (5.18). Then $v$ can be extended by the right–hand side to a radiating solution of (5.17).*

*Proof.* The same arguments used in the proof of Theorem II.10 give the equivalence in this case again. □

By this equivalence result, in order to study solvability of the weak transmission problem we can analysize the IDE. With appropriately defined operators we interpret the IDE as an operator equation:

**Definition V.7.** Let $k > 0$. For $\kappa \in \Pi$ define the linear bounded operators $A_\kappa \colon H^1(D, \mathbb{C}^2) \to H^1(D, \mathbb{C}^2)$ and $B_\kappa \colon H^1(D, \mathbb{C}^2) \to H^1(D, \mathbb{C}^2)$ by

$$(A_\kappa v)(x) := \operatorname{div} \iint_D P(y) \nabla v(y) \Phi_\kappa(x, y) \, dy,$$

$$(B_\kappa v)(x) := k^2 \iint_D Q(y) v(y) \Phi_\kappa(x, y) \, dy$$

for $x \in D$ and the function

$$f(x) = k^2 \iint_D g(y) \, \Phi_k(x, y) \, dy - \operatorname{div} \iint_D h(y) \, \Phi_k(x, y) \, dy$$

for $x \in D$. With these operators the IDE reads

$$(I + A_k - B_k) v = f.$$

**Assumption V.8.** *Let $k > 0$ the wave number. Additionally to Assumption V.2, assume that there exist positive constants $c_1, c_2$ and an angle $\phi \in [0, 2\pi)$ such that $\tilde{A} = \tilde{A}(x)$ and $B = B(x)$ satisfy*

$$\operatorname{Re}\left[e^{i\phi}(\tilde{A}\xi) \cdot \overline{\xi}\right] \geq c_1 |\xi|^2 \quad \text{and} \quad \operatorname{Re}\left[e^{i\phi}(B\xi) \cdot \overline{\xi}\right] \geq c_2 |\xi|^2$$

*on $D$ for almost all $\xi \in \mathbb{C}^2$.*

**Theorem V.9.** *Let Assumption V.8 be satisfied. Then:*

*(a) The operators $A_k - A_i$ and $B_k - B_i$ are compact.*

*(b) The operator $I + A_i - B_i$ is boundedly invertible in $H^1(D, \mathbb{C}^2)$.*

## 2. Direct transmission problem

*Proof.* (a) The operator $B_\kappa$ is compact since $B_\kappa v \in H^2(D, \mathbb{C}^2)$ (Colton and Kress [15]) and the embedding from $H^2$ into $H^1$ is compact (see Adams [2]).

$$(A_k v - A_i v)(x) = \iint_D P(y) v(y) \cdot \nabla_x (\Phi_k - \Phi_i)(x, y) \, dy.$$

Recall the remark to the definition of $\Phi_\kappa$. The singularity is of the form $\log(\kappa |x - y|)$. Hence $\nabla_x \Phi_\kappa$ has a singularity of the form $\frac{x-y}{|x-y|^2}$ and the difference $\nabla_x (\Phi_k - \Phi_i)$ is smooth and $A_k - A_i$ represents a volume potential with a smooth kernel function. It is weakly singular of order 0 and therefore compact (compare Lemma II.15).

(b) For any $f \in H^1(D, \mathbb{C}^2)$ consider the equation $(I + A_i - B_i)u = f$. As in the proof of Theorem II.17 we look at the difference $v := u - f$. $v$ solves $(I + A_i - B_i)v = B_i f - A_i f$ which (by Theorem V.6) is equivalent to

$$\iint_{\mathbb{R}^2} (A \nabla v) \cdot \nabla \psi + (Bv) \cdot \psi \, dx = \iint_D (P \nabla f) \cdot \psi + k^2 (Qf) \cdot \psi \, dx$$

for all $\psi \in H_c^1(\mathbb{R}^2, \mathbb{C}^2)$. Obviously, the left–hand side defines a bounded and coercive sesqui–linear form on $H^1(\mathbb{R}^2, \mathbb{C}^2)$ (note the exponential decay by the IDE) and the right–hand side defines a bounded conjugate–linear form on $H^1(\mathbb{R}^2, \mathbb{C}^2)$. Hence, by the Lax–Milgram lemma there exists a unique solution for every $f \in H^1(D, \mathbb{C}^2)$. Then $u := v|_D + f$ solves our initial equation. □

We proceed with an uniqueness result for the homogeneous problem and apply Fredholm's alternative.

**Assumption V.10.** *Additionally to Assumption V.8 assume that*

$$\operatorname{Im}[(\tilde{A}\xi) \cdot \bar{\xi}] \leq 0 \quad \text{and} \quad \operatorname{Im}[(B\xi) \cdot \bar{\xi}] \geq 0 \quad (5.19)$$

*on $D$ for almost all $\xi \in \mathbb{C}^2$ and let one of the following conditions be satisfied:*

*(a) $\operatorname{Im}[(B\xi) \cdot \bar{\xi}] > 0$ on $D$ for almost all $\xi \in \mathbb{C}^2$,*

*(b) $A \in C^1(\mathbb{R}^2, \mathbb{C}^{4 \times 4})$ and $B \in C^1(\mathbb{R}^2, \mathbb{C}^{2 \times 2})$.*

**Theorem V.11 (Uniqueness).** *Under Assumption V.10 the homogeneous transmission problem — Problem 5 with $g \equiv 0$ and $h \equiv 0$ — has at most one solution.*

*Proof.* Assume that $v$ is a solution of the homogeneous transmission problem and set $\psi = \phi \overline{v}$ in (5.17) where $\phi \in C^\infty(\mathbb{R}^2)$ is some mollifier with $\phi(x) = 1$ for $|x| \leq R$ and $\phi(x) = 0$ for $|x| \geq 2R$. $R$ is chosen such that $|x| < R$ for all $x \in \overline{D}$. Then, by Greens formula

$$0 = \iint_{|x|<R} (A\nabla v) \cdot \nabla \overline{v} - k^2(Bv) \cdot \overline{v} \, dx$$

$$+ \iint_{R<|x|<2R} (A\nabla v) \cdot \nabla(\phi \overline{v}) - k^2(Bv) \cdot (\phi \overline{v}) \, dx \quad (5.20)$$

$$= \iint_{|x|<R} (A\nabla v) \cdot \nabla \overline{v} - k^2(Bv) \cdot \overline{v} \, dx - \int_{|x|=R} \frac{\partial v}{\partial \nu} \cdot \overline{v} \, ds.$$

By the assumptions (5.19), taking the imaginary part of the last equation yields

$$\text{Im} \int_{|x|=R} \frac{\partial v}{\partial \nu} \cdot \overline{v} \, ds \leq 0.$$

We follow Kirsch [22] to show that $v$ vanishes outside of $D$: Using the binomial $|x - iy|^2 = |x|^2 + |y|^2 - 2\text{Im}(x\overline{y})$ we estimate

$$\int_{|x|=R} |\tfrac{\partial v}{\partial \nu} - ikv|^2 \, ds = \int_{|x|=R} |\tfrac{\partial v}{\partial \nu}|^2 + |v|^2 \, ds - 2k\text{Im} \int_{|x|=R} \tfrac{\partial v}{\partial \nu} \cdot \overline{v} \, ds$$

$$\geq \int_{|x|=R} |\tfrac{\partial v}{\partial \nu}|^2 + |v|^2 \, ds.$$

The Sommerfeld radiation condition yields that

$$\lim_{R \to \infty} \int_{|x|=R} |v|^2 \, ds = 0$$

and Rellich's lemma implies that $v$ vanishes in the exterior of $\overline{D}$. See Lemma 3.14 and section 3.10 in Colton and Kress [14].

In case that $A$ and $B$ are smooth we can apply the unique continuation principle (see Lemma 8.5 in Colton and Kress [15]). This yields that $v$ also vanishes in $D$.

In case that (a) from Assumption V.10 is satisfied equation (5.20) now reads
$$\iint_D (A\nabla v)\cdot\nabla\overline{v} - k^2(Bv)\cdot\overline{v}\,dx = 0.$$
Taking the imaginary part yields
$$0 < \operatorname{Im}\iint_D k^2(Bv)\cdot\overline{v}\,dx = \operatorname{Im}\iint_D (A\nabla v)\cdot\nabla\overline{v}\,dx \leq 0$$
by Assumption V.10. Hence $v=0$ in $D$. □

**Corollary V.12.** *Let Assumption V.10 be satisfied. For every source $(g,h)\in L^2(D,\mathbb{C}^2)\times L^2(D,\mathbb{C}^4)$ there exists a unique radiating solution $v\in H^1_{\mathrm{loc}}(\mathbb{R}^2,\mathbb{C}^2)$ of (5.17). Furthermore, for any compact set $B\supset \overline{D}$ there exists a constant $C>0$ such that*
$$\|v\|_{H^1(B,\mathbb{C}^2)} \leq C\,\|(g,h)\|_{L^2(D,\mathbb{C}^2)\times L^2(D,\mathbb{C}^4)}$$
*for all $(g,h)\in L^2(D,\mathbb{C}^2)\times L^2(D,\mathbb{C}^4)$.*

## 3. Factorization Method

In this section we assume that the direct transmission problem is uniquely solvable. It is well known (see Cakoni and Colton [10]) that radiating solutions to the Helmholtz equation admit a far field pattern. In our case
$$v(x) = \frac{e^{ik|x|}}{\sqrt{|x|}}\left\{v^\infty(\hat{x}) + \mathcal{O}\left(\frac{1}{|x|}\right)\right\},\qquad |x|\to\infty,$$
uniformly in all directions $\hat{x}:=x/|x|\in\mathbb{S}^1$. Here, the far field pattern $v^\infty=(v_1^\infty,v_2^\infty)^\top$ is given by
$$v_j^\infty(\hat{x}) = \frac{e^{i\pi/4}}{\sqrt{8\pi k}}\int_{\partial D} v_j(y)\frac{\partial e^{-ik\hat{x}\cdot y}}{\partial \nu(y)} - \frac{\partial v_j}{\partial \nu}(y)e^{-ik\hat{x}\cdot y}\,ds(y),\qquad j=1,2,$$
for $\hat{x}\in\mathbb{S}^1$. As incident fields $v^i=(v_1^i,v_2^i)^\top$ we choose plane waves of the form $v^i(x)=pe^{ikd\cdot x}$ with direction of incidence $d\in\mathbb{S}^1$ and polarization vector $p\in\mathbb{C}^2$. Denote by $v^\infty(\hat{x};d,p)$ the far field pattern resulting from a plane wave with direction of incidence $d$ and polarization $p$. Now we can formulate the inverse problem:

**Problem 6 (Inverse problem).** Given the wave number $k > 0$ and the far field patterns $v^\infty(\hat{x}; d, p)$ for all $\hat{x}, d \in \mathbb{S}^1$ and $p \in \mathbb{C}^2$ determine the shape of the scattering obstacle $D$.

$v^\infty(\,\cdot\,; d, p)$ depends linearly on $p$. Hence, we define the linear FAR FIELD OPERATOR $\mathcal{F} \colon L^2(\mathbb{S}^1, \mathbb{C}^2) \to L^2(\mathbb{S}^1, \mathbb{C}^2)$ by

$$(\mathcal{F}p)(\hat{x}) := \int_{\mathbb{S}^1} v^\infty\bigl(\hat{x}; d, p(d)\bigr)\, \mathrm{d}s(d), \quad \hat{x} \in \mathbb{S}^1.$$

$\mathcal{F}p$ is the far field pattern which corresponds to the incident field

$$v_p^i(x) = \int_{\mathbb{S}^1} p(d) e^{ikd\cdot x}\, \mathrm{d}s(d), \quad x \in \mathbb{R}^2.$$

Furthermore, define the HERGLOTZ OPERATOR $\mathcal{H}$ from $L^2(\mathbb{S}^1, \mathbb{C}^2)$ into $L^2(D, \mathbb{C}^2) \times L^2(D, \mathbb{C}^4)$ with $\mathcal{H}p = (\mathcal{H}_1 p, \mathcal{H}_2 p)^\top$ by

$$(\mathcal{H}_1 p)(y) := \int_{\mathbb{S}^1} p(d) e^{ikd\cdot y}\, \mathrm{d}s(d)$$

and

$$(\mathcal{H}_2 p)(y) := \mathrm{grad} \int_{\mathbb{S}^1} p(d) e^{ikd\cdot y}\, \mathrm{d}s(d)$$

for $y \in D$. More precisely, with $p = (p_1, p_2)^\top \in L^2(\mathbb{S}^1, \mathbb{C}^2)$

$$(\mathcal{H}_2 p)(y) = \left( \mathrm{grad} \int_{\mathbb{S}^1} p_1(d) e^{ikd\cdot y}\, \mathrm{d}s(d), \mathrm{grad} \int_{\mathbb{S}^1} p_2(d) e^{ikd\cdot y}\, \mathrm{d}s(d) \right)^\top.$$

The adjoint operator $\mathcal{H}^* \colon L^2(D, \mathbb{C}^2) \times L^2(D, \mathbb{C}^4) \to L^2(\mathbb{S}^1, \mathbb{C}^2)$ is given by $\mathcal{H}^* \varphi = \mathcal{H}_1^* \varphi_1 + \mathcal{H}_2^* \varphi_2$ for $\varphi = (\varphi_1, \varphi_2)^\top$ with

$$(\mathcal{H}_1^* \varphi_1)(d) = \iint_D \varphi_1(y) e^{-ikd\cdot y}\, \mathrm{d}y$$

and

$$(\mathcal{H}_2^* \varphi_2)(d) = -ik\, d \cdot \iint_D \varphi_2(y) e^{-ikd\cdot y}\, \mathrm{d}y.$$

More precisely, with $\varphi_2 = (\varphi_2^1, \varphi_2^2, \varphi_2^3, \varphi_2^4)^\top \in L^2(D, \mathbb{C}^4)$

$$(\mathcal{H}_2^* \varphi_2)(d) = -ik \begin{pmatrix} d \cdot \iint_D \begin{pmatrix} \varphi_2^1(y) \\ \varphi_2^2(y) \end{pmatrix} e^{-ikd\cdot y}\, \mathrm{d}y \\ d \cdot \iint_D \begin{pmatrix} \varphi_2^3(y) \\ \varphi_2^4(y) \end{pmatrix} e^{-ikd\cdot y}\, \mathrm{d}y \end{pmatrix}.$$

## 3. Factorization Method

Using the far field pattern of $\Phi_k$ we see that $\mathcal{H}^*\varphi = w^\infty/\gamma$ where the constant $\gamma := \frac{e^{i\pi/4}}{\sqrt{8\pi k}}$ and

$$w(x) = \iint_D \varphi_1(y)\,\Phi_k(x,y)\,\mathrm{d}y - \mathrm{div}\iint_D \varphi_2(y)\,\Phi_k(x,y)\,\mathrm{d}y$$

is a radiating solution of

$$\iint_{\mathbb{R}^2} \nabla w \cdot \nabla \psi - k^2 w \cdot \psi\,\mathrm{d}x = \iint_D \varphi_1 \cdot \psi + \varphi_2 \cdot \nabla \psi\,\mathrm{d}x$$

for all $\psi \in H_c^1(\mathbb{R}^2, \mathbb{C}^2)$. Introduce the DATA–TO–PATTERN OPERATOR $G \colon L^2(D,\mathbb{C}^2) \times L^2(D,\mathbb{C}^4) \to L^2(\mathbb{S}^1, \mathbb{C}^2)$, $f \mapsto v^\infty$ where $v$ is the radiating solution of

$$\iint_{\mathbb{R}^2}(A\nabla v)\cdot \nabla \psi - k^2(Bv)\cdot \psi\,\mathrm{d}x = \iint_D k^2(Qf_1)\cdot \psi + (Pf_2)\cdot \nabla \psi\,\mathrm{d}x \quad (5.21)$$

for all $\psi \in H_c^1(\mathbb{R}^2, \mathbb{C}^2)$ which is equivalent to

$$\iint_{\mathbb{R}^2} \nabla v \cdot \nabla \psi - k^2 v \cdot \psi\,\mathrm{d}x = \iint_D k^2(Qw_1)\cdot \psi + (Pw_2)\cdot \nabla \psi\,\mathrm{d}x \quad (5.22)$$

for all $\psi \in H_c^1(\mathbb{R}^2, \mathbb{C}^2)$ with $w_1 := f_1 + v$ and $w_2 := f_2 + \nabla v$. Then $\mathcal{F} = G\mathcal{H}$. Choose $\varphi = \mathcal{T}f$ with

$$\mathcal{T}f = \begin{pmatrix} k^2 Q(f_1+v) \\ P(f_2+\nabla v) \end{pmatrix}$$

where $v$ is the radiating solution of (5.21). Then $Gf = \gamma \mathcal{H}^* \mathcal{T} f$ and we have the factorization

$$\mathcal{F} = \gamma \mathcal{H}^* \mathcal{T} \mathcal{H}.$$

**Remark V.13.** Note that $\mathcal{T}$ is injective. Indeed, if $\mathcal{T}f = 0$ then $w_1 = 0$ and $w_2 = 0$ and – by equation (5.22) – $v$ is a radiating solution of the Helmholtz equation in $\mathbb{R}^2$. The unique solution is $v \equiv 0$. Hence, $f$ also vanishes.

We continue with a characterization of $D$ by the range of $\mathcal{H}^*$.

$$L^2(\mathbb{S}^1,\mathbb{C}^2) \xrightarrow{\gamma^{-1}\mathcal{F}} L^2(\mathbb{S}^1,\mathbb{C}^2)$$

$$\mathcal{H}\downarrow \quad /// \quad \uparrow \mathcal{H}^*$$

$$L^2(D,\mathbb{C}^2)\times L^2(D,\mathbb{C}^4) \xrightarrow{\mathcal{T}} L^2(D,\mathbb{C}^2)\times L^2(D,\mathbb{C}^4)$$

Figure V.2: Factorization of $\mathcal{F}$.

**Theorem V.14.** *For any $z \in \mathbb{R}^3$ and fixed $p_1 \in \mathbb{C}^2$ and $p_2 \in \mathbb{C}^4$ define $\phi_z \in L^2(\mathbb{S}^1,\mathbb{C}^2)$ by*

$$\phi_z(d) := \{p_1 - ik\, d\cdot p_2\} e^{-ik\, d\cdot z}, \qquad d \in \mathbb{S}^1. \tag{5.23}$$

*Then $z \in D$ if, and only if, $\phi_z \in \mathcal{R}(\mathcal{H}^*)$.*

Here again, with $p_2 = (p_2^1, p_2^2, p_2^3, p_2^4)^\top \in \mathbb{C}^4$ we define

$$d\cdot p_2 := \begin{pmatrix} d\cdot \begin{pmatrix} p_2^1 \\ p_2^2 \end{pmatrix} \\ d\cdot \begin{pmatrix} p_2^3 \\ p_2^4 \end{pmatrix} \end{pmatrix} \in \mathbb{C}^2.$$

*Proof.* Let $z \in D$. We apply Lemma III.29 with $q = 1$. Analogously to Theorem III.30 there exist $\varphi_1, \varphi_2 \in L^2(D)$ such that $\phi_z = \mathcal{H}^*(p_1\varphi_1, p_2\varphi_2)$. □

**Theorem V.15 (Absorbing media).** *Assume that there exist two positive constants $c_P$ and $c_Q$ such that the contrasts $\tilde{P} = \tilde{P}(x)$, $Q = Q(x)$ satisfy*

$$\mathrm{Im}\left[(\tilde{P}\xi)\cdot \overline{\xi}\right] \geq c_P |\xi|^2 \quad \text{and} \quad \mathrm{Im}\left[(Q\xi)\cdot \overline{\xi}\right] \geq c_Q |\xi|^2$$

*on $D$ for almost all $\xi \in \mathbb{C}^2$. Then:*

(a) *The operator $\mathrm{Im}\,\mathcal{T} = \frac{1}{2i}(\mathcal{T}-\mathcal{T}^*)$ is coercive on $\mathcal{R}(\mathcal{H})$; that is, there exists a positive constant $c$ such that*

$$\mathrm{Im}\,(\mathcal{T}f, f) \geq c\|f\|^2 \quad \text{for all } f \in \mathcal{R}(\mathcal{H}).$$

## 3. Factorization Method

*(b) The ranges of* $\operatorname{Im}(\gamma^{-1}\mathcal{F})^{1/2}$ *and* $\mathcal{H}^*$ *coincide.*

*Proof.* As in the proof of Theorem III.19 we deduce an expression for $(\mathcal{T}f, f)$ involving the far field pattern of $v$. Abbreviating $w_1 = f_1 + v$ and $w_2 = f_2 + \nabla v$ we have $\mathcal{T}f = (k^2 Q w_1, P w_2)^\top$. $v$ solves equation (5.22). In this equation, we choose as test function $\psi = \phi \overline{v}$ with the cutoff function $\phi$ used in the proof of III.19. Then,

$$(\mathcal{T}f, f)_{L^2} = \left( \begin{pmatrix} k^2 Q w_1 \\ P w_2 \end{pmatrix}, \begin{pmatrix} w_1 - v \\ w_2 - \nabla v \end{pmatrix} \right)_{L^2}$$

$$= \iint_D k^2 (Q w_1) \cdot \overline{w_1} + (P w_2) \cdot \overline{w_2} \, dx$$

$$\quad - \iint_D k^2 (Q w_1) \cdot \overline{v} + (P w_2) \cdot \nabla \overline{v} \, dx$$

$$= \iint_D k^2 (Q w_1) \cdot \overline{w_1} + (P w_2) \cdot \overline{w_2} \, dx$$

$$\quad - \iint_{|x|<R} |\nabla v|^2 - k^2 |v|^2 \, dx + \int_{|x|=R} \frac{\partial v}{\partial \nu} \overline{v} \, ds.$$

From the Sommerfeld radiation condition and the far field expansion we conclude that

$$\lim_{R \to \infty} \int_{|x|=R} \frac{\partial v}{\partial \nu} \overline{v} \, ds = ik \int_{\mathbb{S}^1} |v^\infty|^2 \, ds.$$

Hence, letting $R$ tend to infinity:

$$\operatorname{Im}(\mathcal{T}f, f) = \iint_D k^2 \operatorname{Im}\left[(Q w_1) \cdot \overline{w_1}\right] + \operatorname{Im}\left[(P w_2) \cdot \overline{w_2}\right] dx + k \int_{\mathbb{S}^1} |v^\infty|^2 \, ds$$

The coercivity of $\operatorname{Im} \mathcal{T}$ follows with the arguments used in part (b) of the proof III.19. □

**Corollary V.16.** *$\operatorname{Im} \mathcal{F}$ admits a complete eigensystem $\{\lambda_j, \psi_j\}_{j \in \mathbb{N}}$ with positive eigenvalues $\lambda_j$. For $z \in \mathbb{R}^2$ define $\phi_z$ by (5.23). Then $z \in D$ if, and only if, $\phi_z \in \mathcal{R}\left((\operatorname{Im} \mathcal{F})^{1/2}\right)$. Or equivalently,*

$$z \in D \iff W(z) := \left[ \sum_{j \in \mathbb{N}} \frac{|(\phi_z, \psi_j)|^2}{\lambda_j} \right]^{-1} > 0$$

For the general case we need to decompose the middle operator $\mathcal{T}$ as sum of a coercive and a compact one. Therefore, define $\mathcal{T}_0$ by

$$\mathcal{T}_0 f := \begin{pmatrix} k^2 Q(f_1 + v_0) \\ P(f_2 + \nabla v_0) \end{pmatrix}$$

where $v_0$ is the radiating solution of

$$\iint_{\mathbb{R}^2} (A\nabla v_0) \cdot \nabla \psi + (Bv_0) \cdot \psi \, dx = \iint_D k^2 (Qf_1) \cdot \psi + (Pf_2) \cdot \nabla \psi \, dx \quad (5.24)$$

for all $\psi \in H_c^1(\mathbb{R}^2, \mathbb{C}^2)$. Under Assumption V.8 this equation is uniquely solvable and using the IDE we see that $v_0$ decays exponentially. With this operator we decompose the middle operator as $\mathcal{T} = \mathcal{T}_0 + (\mathcal{T} - \mathcal{T}_0)$ and show in part (b) and (c) of the following theorem that $\mathcal{T}_0$ is coercive and that the difference is compact.

**Theorem V.17 (General case).** *Assume that*

$$\operatorname{Im}\left[(\tilde{P}\xi) \cdot \overline{\xi}\right] \geq 0 \quad \text{and} \quad \operatorname{Im}\left[(Q\xi) \cdot \overline{\xi}\right] \geq 0$$

*on $D$ for almost all $\xi \in \mathbb{C}^2$. Furthermore, assume that there exist two positive constants $c_P$, $c_Q$ and an angle $\phi \in [0, 2\pi)$ such that*

$$\operatorname{Re}\left[e^{i\phi}(\tilde{P}\xi) \cdot \overline{\xi}\right] \geq c_P |\xi|^2 \quad \text{and} \quad \operatorname{Re}\left[e^{i\phi}(Q\xi) \cdot \overline{\xi}\right] \geq c_Q |\xi|^2$$

*on $D$ for almost all $\xi \in \mathbb{C}^2$ and*

$$\operatorname{Re}\left[e^{i\phi}(\tilde{A}\xi) \cdot \overline{\xi}\right] \geq 0 \quad \text{and} \quad \operatorname{Re}\left[e^{i\phi}(B\xi) \cdot \overline{\xi}\right] \geq 0$$

*on $D$ for almost all $\xi \in \mathbb{C}^2$. Define the operator $\mathcal{F}_\#$ by*

$$\mathcal{F}_\# := |\operatorname{Re}(\gamma^{-1} e^{i\phi} \mathcal{F})| + \operatorname{Im}(\gamma^{-1} \mathcal{F}).$$

*Then:*

*(a) $\operatorname{Im} \mathcal{T}$ is non–negative; that is, for all $f \in L^2(D, \mathbb{C}^2) \times L^2(D, \mathbb{C}^4)$: $\operatorname{Im}(\mathcal{T}f, f) \geq 0$.*

*(b) $\mathcal{T}_0$ is coercive; that is, there exists a constant $c > 0$ such that*

$$\operatorname{Re}(e^{i\phi} \mathcal{T}_0 f, f) \geq c \|f\|^2 \quad \text{for all } f \in \mathcal{R}(\mathcal{H}).$$

## 3. Factorization Method

(c) The operator $\mathcal{T} - \mathcal{T}_0$ is compact from $L^2(D,\mathbb{C}^2) \times L^2(D,\mathbb{C}^4)$ into itself.

(d) $\mathcal{F}_\#$ is injective and the ranges of $\mathcal{F}_\#^{1/2}$ and $\mathcal{H}^*$ conincide.

*Proof.* (a) From the previous theorem we already know that

$$\mathrm{Im}\,(\mathcal{T}f,f) = \iint_D k^2 \mathrm{Im}\left[(Qw_1)\cdot\overline{w_1}\right] + \mathrm{Im}\left[(Pw_2)\cdot\overline{w_2}\right]\,\mathrm{d}x + k\int_{\mathbb{S}^1}|v^\infty|^2\,\mathrm{d}s.$$

Hence, under the positivity assumptions for the imaginary parts of the contrasts $P$ and $Q$: $\mathrm{Im}\,(\mathcal{T}f,f) \geq 0$.

(b) Using $\psi = \overline{v}_0$ in (5.24) (note that $v_0$ decays exponentially) we deduce

$$(\mathcal{T}_0 f, f) =$$
$$= \iint_D k^2(Qf_1)\cdot\overline{f_1} + (Pf_2)\cdot\overline{f_2}\,\mathrm{d}x + \iint_D k^2(Qf_1)\cdot\overline{v}_0 + (Pf_2)\cdot\nabla\overline{v_0}\,\mathrm{d}x$$
$$= \iint_D k^2(Qf_1)\cdot\overline{f_1} + (Pf_2)\cdot\overline{f_2}\,\mathrm{d}x + \iint_{\mathbb{R}^2} (A\nabla v_0)\cdot\nabla\overline{v_0} + k^2(Bv_0)\cdot\overline{v_0}\,\mathrm{d}x.$$

Hence,

$$\mathrm{Re}\,(e^{i\phi}\mathcal{T}_0 f, f) \geq \iint_D k^2 \mathrm{Re}\left[e^{i\phi}(Qf_1)\cdot\overline{f_1}\right] + \mathrm{Re}\left[e^{i\phi}(Pf_2)\cdot\overline{f_2}\right]\,\mathrm{d}x$$
$$\geq \min\{k^2 c_Q, c_P\}\left\{\|f_1\|^2_{L^2(D,\mathbb{C}^2)} + \|f_2\|^2_{L^2(D,\mathbb{C}^4)}\right\}.$$

(c) $(\mathcal{T} - \mathcal{T}_0)f = (k^2 Q(v - v_0), P(\nabla v - \nabla v_0))^\top$ and the difference $w := v - v_0$ is a radiating solution of

$$\iint_{\mathbb{R}^2}(A\nabla w)\cdot\nabla\psi - k^2(Bw)\cdot\psi\,\mathrm{d}x = (k^2+1)\iint_{\mathbb{R}^2}(Bv_0)\cdot\psi\,\mathrm{d}x \quad (5.25)$$

for all $\psi \in H^1_c(\mathbb{R}^2,\mathbb{C}^2)$. Since $v, v_0 \in H^1_{\mathrm{loc}}(\mathbb{R}^2,\mathbb{C}^2)$: $w|_D \in H^1(D,\mathbb{C}^2)$ which is compactly embedded in $L^2(D,\mathbb{C}^2)$. Hence the first component of $\mathcal{T} - \mathcal{T}_0$ is compact. To show compactness of the second component we repeat the procedure from the proof of part (b) of Theorem III.25. Starting with a sequence $(f^n)_n \subset L^2(D,\mathbb{C}^2) \times L^2(D,\mathbb{C}^4)$ converging weakly to zero, the unique solvability of our transmission problem yields that the corresponding sequences of solutions $(v^n)_n, (v_0^n)_n$ converge weakly to zero in $H^1(K,\mathbb{C}^2)$ for any ball $K$. Then they converge in the norm of $L^2$ to zero (by compact embedding). Let $w^n = v^n - v_0^n$. We choose two balls $K_1, K_2$ with $K_1 \supset K_2 \supset \overline{D}$ and $\psi^n := \phi \overline{w^n}$ in equation (5.25) where $\phi \in \mathcal{C}^\infty$ is a cutoff function with $\phi \equiv 1$ in $\overline{K_2}$ and $\phi \equiv 0$ in $\mathbb{R}^3 \setminus K_1$. Then (5.25) reads

$$\iint_{K_2} (A\nabla w^n) \cdot \nabla \overline{w^n} - k^2(Bw^n) \cdot \overline{w^n}\, dx =$$

$$-\iint_{K_1 \setminus K_2} \nabla w^n \cdot \nabla(\phi \overline{w^n}) - k^2 w^n \cdot (\phi \overline{w^n})\, dx + (k^2+1) \iint_{K_1} (Bv_0^n) \cdot (\phi \overline{w^n})\, dx$$

$$= \int_{\partial K_2} \frac{\partial w^n}{\partial \nu} \cdot \overline{w^n}\, ds + (k^2+1) \iint_{K_2} (Bv_0^n) \cdot (\phi \overline{w^n})\, dx$$

by Green's Theorem (since $\Delta w^n + k^2 w^n = 0$ in $\mathbb{R}^2 \setminus \overline{D}$). Now we estimate:

$$|r.h.s.| \leq \left| \int_{\partial K_2} \frac{\partial w^n}{\partial \nu} \cdot \overline{w^n}\, ds \right| + C \|v_0^n\| \|w^n\| \xrightarrow{n \to \infty} 0.$$

Hence,

$$\iint_{K_2} (A\nabla w^n) \cdot \nabla \overline{w^n} - k^2(Bw^n) \cdot \overline{w^n}\, dx \xrightarrow{n \to \infty} 0$$

which implies $\|\nabla w^n\| \to 0$ as $n \to \infty$.

(d) This assertion is an application of the range identity result in Theorem III.32. □

**Corollary V.18.** $\mathcal{F}_\#$ *admits a complete eigensystem* $\{\lambda_j, \psi_j\}_{j \in \mathbb{N}}$ *with positive eigenvalues* $\lambda_j$. *For* $z \in \mathbb{R}^2$ *define* $\phi_z$ *by* (5.23). *Then* $z \in D$ *if, and only if,* $\phi_z \in \mathcal{R}\left(\mathcal{F}_\#^{1/2}\right)$. *Or equivalently,*

$$z \in D \iff W(z) := \left[ \sum_{j \in \mathbb{N}} \frac{|(\phi_z, \psi_j)|^2}{\lambda_j} \right]^{-1} > 0.$$

4. Application: Scattering by a chiral cylinder 125

|  |  |  |  |
|---|---|---|---|
| DP | exist. | Re $\left[e^{i\phi}(\tilde{A}x)\cdot\overline{x}\right] \geq c_A|x|^2$, | Re $\left[e^{i\phi}(Bx)\cdot\overline{x}\right] \geq c_B|x|^2$ |
|  | uniq. | Im $\left[(\tilde{A}x)\cdot\overline{x}\right] \leq 0$, | Im $\left[(Bx)\cdot\overline{x}\right] \geq (>)0$ |
| IP | absorb. | Im $\left[(\tilde{P}x)\cdot\overline{x}\right] \geq c_P|x|^2$, | Im $\left[(Qx)\cdot\overline{x}\right] \geq c_Q|x|^2$ |
|  | general | Re $\left[e^{i\phi}(\tilde{P}x)\cdot\overline{x}\right] \geq c_P|x|^2$, | Re $\left[e^{i\phi}(Qx)\cdot\overline{x}\right] \geq c_Q|x|^2$ |
|  |  | Im $\left[(\tilde{P}x)\cdot\overline{x}\right] \geq 0$, | Im $\left[(Qx)\cdot\overline{x}\right] \geq 0$ |

Table V.1: Overview: Assumptions on the matrices and the contrasts.

## 4. Application: Scattering by a chiral cylinder

In the first section we developed a transmission problem modeling the scattering by a chiral cylinder. We found the vector Helmholtz equation

$$\operatorname{div}(A\nabla v) + k^2 B v = 0$$

for $v = (v_1, v_2)^\top$. The direct and inverse problem were treated in sections 2 and 3. Now, we apply the general results. The matrices are given by

$$A = \begin{pmatrix} \frac{1}{\mu} & 0 & -ik\beta & 0 \\ 0 & \frac{1}{\mu} & 0 & -ik\beta \\ \hline ik\beta & 0 & \frac{1}{\varepsilon} & 0 \\ 0 & ik\beta & 0 & \frac{1}{\varepsilon} \end{pmatrix}$$

and

$$B = \frac{1}{1 - k^2 \varepsilon \mu \beta^2} \begin{pmatrix} \varepsilon & ik\varepsilon\mu\beta \\ -ik\varepsilon\mu\beta & \mu \end{pmatrix} = \begin{pmatrix} \frac{\varepsilon}{\vartheta} & \frac{i}{k\beta}(\frac{1}{\vartheta} - 1) \\ -\frac{i}{k\beta}(\frac{1}{\vartheta} - 1) & \frac{\mu}{\vartheta} \end{pmatrix}.$$

where we introduced $\vartheta := 1 - k^2 \varepsilon \mu \beta^2$.

In what follows, we analyze the conditions for the material parameters $\varepsilon, \mu$ and $\beta$ such that the matrices $A$ and $B$ and the contrasts $P = I - A$ and $Q = B - I$ satisfy the assumptions used in the previous sections. Table V.1 recalls the main assumptions. Given a matrix $M \in \mathbb{C}^{2\times 2}$ of the form

$$M = \begin{pmatrix} a & ib \\ -ib & c \end{pmatrix}$$

with complex numbers $a, b, c \in \mathbb{C}$ the following can be easily computed for $x = (x_1, x_2)^\top \in \mathbb{C}^2$:

$$(Mx) \cdot \overline{x} = a|x_1|^2 + 2b \operatorname{Im}(x_1\overline{x_2}) + c|x_2|^2,$$

$$\operatorname{Re}[(Mx) \cdot \overline{x}] = \operatorname{Re} a \left| x_1 + i \tfrac{\operatorname{Re} b}{\operatorname{Re} a} x_2 \right|^2 + \operatorname{Re} c \left( 1 - \tfrac{(\operatorname{Re} b)^2}{\operatorname{Re} a \operatorname{Re} c} \right) |x_2|^2, \quad (5.26)$$

$$\operatorname{Im}[(Mx) \cdot \overline{x}] = \operatorname{Im} a \left| x_1 + i \tfrac{\operatorname{Im} b}{\operatorname{Im} a} x_2 \right|^2 + \operatorname{Im} c \left( 1 - \tfrac{(\operatorname{Im} b)^2}{\operatorname{Im} a \operatorname{Im} c} \right) |x_2|^2. \quad (5.27)$$

We give an example of possible values for $\varepsilon, \mu$ and $\beta$ such that the assumptions on $\tilde{A}, B$ and $\tilde{P}, Q$ are satisfied:

**Proposition V.19 (Direct problem).** *Given the wave number $k > 0$. Assume that there exist radii $0 < r_1 < 1 < r_2$ and angles $0 < \varphi_2 < \phi < \pi/4$ such that*

$$\varepsilon, \mu \in \left\{ re^{i\varphi} : r \in [r_1, r_2], \varphi \in [0, \varphi_2] \right\}$$

*almost everywhere. For a sufficiently small positive constant $\lambda$ such that*

$$k^2 \beta^2 \leq \lambda$$

*almost everywhere, there exist positive constants $c_A, c_B$ such that the coercivity conditions*

$$\operatorname{Re}\left[ e^{-i\phi}(\tilde{A}x) \cdot \overline{x} \right] \geq c_A |x|^2 \quad \text{and} \quad \operatorname{Re}\left[ e^{-i\phi}(Bx) \cdot \overline{x} \right] \geq c_B |x|^2$$

*are fullfilled and, furthermore,*

$$\operatorname{Im}\left[(\tilde{A}x) \cdot \overline{x}\right] \leq 0 \quad \text{and} \quad \operatorname{Im}\left[(Bx) \cdot \overline{x}\right] \geq 0.$$

*Proof.* (i) From (5.26) we conclude that

$$\operatorname{Re}(\tilde{A}x) \cdot \overline{x} = \operatorname{Re} \tfrac{1}{\mu} \left| x_1 - i \tfrac{k\beta}{\operatorname{Re}(\mu^{-1})} x_2 \right|^2 + \operatorname{Re} \tfrac{1}{\varepsilon} \left( 1 - \tfrac{k^2 \beta^2}{\operatorname{Re}(\mu^{-1})\operatorname{Re}(\varepsilon^{-1})} \right) |x_2|^2.$$

By our assumptions the reciprocals $1/\mu$ and $1/\varepsilon$ are contained in the following set of complex numbers

$$\left\{ re^{i\varphi} : r \in [r_2^{-1}, r_1^{-1}], \varphi \in [-\varphi_2, 0] \right\}$$

and $-\varphi_2 > -\phi > -\pi/4$. The numbers of this set have strictly positive real parts and this is still valid after rotation by the angle $-\phi$. Finally

$$\frac{k^2 \beta^2 \cos^2(\phi)}{\operatorname{Re}(e^{-i\phi}\mu^{-1})\operatorname{Re}(e^{-i\phi}\varepsilon^{-1})} = \mathcal{O}(k^2 \beta^2).$$

## 4. Application: Scattering by a chiral cylinder

Therefore, there exists a constant $\lambda$ such that

$$1 - \frac{k^2\beta^2 \cos^2(\phi)}{\operatorname{Re}(e^{-i\phi}\mu^{-1})\operatorname{Re}(e^{-i\phi}\varepsilon^{-1})} > 0$$

for $k^2\beta^2 < \lambda$.

(ii) Define the contrast $q_1 := 1/\vartheta - 1$. From (5.26) we conclude that

$$\operatorname{Re}(Bx) \cdot \overline{x} =$$

$$= \operatorname{Re}\frac{\varepsilon}{\vartheta}\left|x_1 + i\frac{\operatorname{Re} q_1}{k\beta \operatorname{Re}\frac{\varepsilon}{\vartheta}}x_2\right|^2 + \operatorname{Re}\frac{\mu}{\vartheta}\left(1 - \frac{(\operatorname{Re} q_1)^2}{k^2\beta^2 \operatorname{Re}\frac{\varepsilon}{\vartheta}\operatorname{Re}\frac{\mu}{\vartheta}}\right)|x_2|^2.$$

We restrict $k^2\beta^2$ by a positive constant $\lambda$ such that

$$k^2\beta^2 \leq \lambda < r_2^{-2}.$$

By our assumptions on $\varepsilon$ and $\mu$ we compute:

$$\varepsilon\mu \in \{re^{i\varphi} : r \in [r_1^2, r_2^2], \varphi \in [0, 2\varphi_2]\},$$

$$1 - k^2\beta^2\varepsilon\mu \in \{1 + re^{i\varphi} : r \in (0, \lambda r_2^2] \subset (0,1), \varphi \in [-\pi, -\pi + 2\varphi_2]\},$$

$$\frac{1}{1 - k^2\beta^2\varepsilon\mu} \in \{z \in \mathbb{C} : \operatorname{Re}(z) > 1/2, \operatorname{Im} z \geq 0, |z| \leq (1 - \lambda r_2^2)^{-1}\}.$$

We conclude that the values of $\frac{\varepsilon}{1-k^2\beta^2\varepsilon\mu}$ and $\frac{\mu}{1-k^2\beta^2\varepsilon\mu}$ are contained in the set $\{re^{i\varphi} : r \in [r_1/2, r_2(1 - \lambda r_2^2)^{-1}], \varphi \in [0, \pi/2 + \phi)\}$. After rotation by the angle $-\phi$ the numbers of this set have strictly positive real parts. As in the first case we see that

$$\frac{\operatorname{Re}(e^{-i\phi}q_1)^2}{k^2\beta^2 \operatorname{Re}(e^{-i\phi}\frac{\varepsilon}{\vartheta})\operatorname{Re}(e^{-i\phi}\frac{\mu}{\vartheta})} = \mathcal{O}(k^2\beta^2)$$

and by the same argument we can restrict $k^2\beta^2$ such that

$$1 - \frac{\operatorname{Re}(e^{-i\phi}q_1)^2}{k^2\beta^2 \operatorname{Re}(e^{-i\phi}\frac{\varepsilon}{\vartheta})\operatorname{Re}(e^{-i\phi}\frac{\mu}{\vartheta})} > 0.$$

Now we study the uniqueness conditions:

(iii) From (5.27) we deduce

$$\operatorname{Im}(\tilde{A}x) \cdot \overline{x} = \operatorname{Im}\frac{1}{\mu}|x_1|^2 + \operatorname{Im}\frac{1}{\varepsilon}|x_2|^2.$$

Hence, $\mathrm{Im}\,(\tilde{A}x)\cdot\overline{x} \leq 0$ since $\mathrm{Im}\,\varepsilon, \mathrm{Im}\,\mu \geq 0$ by our assumptions.
(iv) From (5.27) we deduce

$$\mathrm{Im}\,(Bx)\cdot\overline{x} =$$
$$= \mathrm{Im}\,\frac{\varepsilon}{\vartheta}\left|x_1 + i\frac{\mathrm{Im}\,q_1}{k\beta\,\mathrm{Im}\,\frac{\varepsilon}{\vartheta}}x_2\right|^2 + \mathrm{Im}\,\frac{\mu}{\vartheta}\left(1 - \frac{(\mathrm{Im}\,q_1)^2}{k^2\beta^2\mathrm{Im}\,\frac{\varepsilon}{\vartheta}\,\mathrm{Im}\,\frac{\mu}{\vartheta}}\right)|x_2|^2.$$

We have already seen that $\mathrm{Im}\,(\varepsilon/\vartheta), \mathrm{Im}\,(\mu/\vartheta) \geq 0$. Again

$$\frac{(\mathrm{Im}\,q_1)^2}{k^2\beta^2\mathrm{Im}\,\frac{\varepsilon}{\vartheta}\,\mathrm{Im}\,\frac{\mu}{\vartheta}} = \mathcal{O}(k^2\beta^2)$$

and we can restrict $k^2\beta^2$ such that

$$1 - \frac{(\mathrm{Im}\,q_1)^2}{k^2\beta^2\mathrm{Im}\,\frac{\varepsilon}{\vartheta}\,\mathrm{Im}\,\frac{\mu}{\vartheta}} > 0.$$

$\square$

**Proposition V.20 (Absorbing media).** *Let the wave number $k > 0$ be given. Assume that*

$$\varepsilon, \mu \in \left\{re^{i\phi} : r \in [r_1, r_2], \varphi \in [\varphi_1, \varphi_2]\right\}$$

*a. e. with radii $0 < r_1 < 1 < r_2$ and angles $0 < \varphi_1 < \varphi_2 < \phi < \pi/4$. For a sufficiently small positive constant $\lambda$ such that $k^2\beta^2 < \lambda$ almost everywhere there exist positive constants $c_P$, $c_Q$ such that the contrasts satisfy*

$$\mathrm{Im}\left[(\tilde{P}x)\cdot\overline{x}\right] \geq c_P|x|^2 \quad\text{and}\quad \mathrm{Im}\left[(Qx)\cdot\overline{x}\right] \geq c_Q|x|^2.$$

*Proof.* (i) From (5.27) we deduce

$$\mathrm{Im}\,(\tilde{P}x)\cdot\overline{x} = \tfrac{\mathrm{Im}\,\mu}{|\mu|^2}|x_1|^2 + \tfrac{\mathrm{Im}\,\varepsilon}{|\varepsilon|^2}|x_2|^2.$$

Hence, $\tilde{P}$ satisfies the above condition if there exists a constant $c_P$ such that $\mathrm{Im}\,\mu \geq c_P|\mu|^2$. We compute

$$\mathrm{Im}\,\mu \geq c_P|\mu|^2 \iff \left(\mathrm{Im}\,\mu - \tfrac{1}{2c_P}\right)^2 + \left(\mathrm{Re}\,\mu\right)^2 \leq \tfrac{1}{4c_P^2}.$$

## 4. Application: Scattering by a chiral cylinder

The values of $\mu$ must be contained in a circle with radius $1/(2c_P)$ and center $i/(2c_P)$. The circle is getting bigger with smaller values of $c_P$ and its boundary always contains the origin. Therefore, it is possible to choose $c_P$ sufficiently small such that the set $\{re^{i\varphi} : r \in [r_1, r_2], \varphi \in [\varphi_1, \varphi_2]\}$ is contained in the circle. Analogously for $\varepsilon$.

(ii) Recall $q_1 = 1/\vartheta - 1$ and define the contrasts $q_\varepsilon := \varepsilon/\vartheta - 1$ and $q_\mu := \mu/\vartheta - 1$. From (5.27) we compute

$$\operatorname{Im}(Qx) \cdot \bar{x} =$$

$$= \operatorname{Im} q_\varepsilon \left| x_1 + i\frac{\operatorname{Im} q_1}{k\beta \operatorname{Im} q_\varepsilon} x_2 \right|^2 + \operatorname{Im} q_\mu \left(1 - \frac{(\operatorname{Im} q_1)^2}{k^2 \beta^2 \operatorname{Im} q_\varepsilon \operatorname{Im} q_\mu}\right) |x_2|^2.$$

First we note that $\operatorname{Im} q_\varepsilon = \operatorname{Im} \frac{\varepsilon}{\vartheta}$ and $\operatorname{Im} q_\mu = \operatorname{Im} \frac{\mu}{\vartheta}$. Under the assumptions in Proposition V.19 we have seen that $\operatorname{Im} \frac{\varepsilon}{\vartheta}, \operatorname{Im} \frac{\mu}{\vartheta} \geq 0$. We easily check that the imaginary parts are bounded away from zero if we start with strictly positive imaginary parts of $\varepsilon$ and $\mu$. Finally,

$$\frac{(\operatorname{Im} q_1)^2}{k^2 \beta^2 \operatorname{Im} q_\varepsilon \operatorname{Im} q_\mu} = \mathcal{O}(k^2 \beta^2)$$

and we can restrict $k^2 \beta^2$ such that

$$1 - \frac{(\operatorname{Im} q_1)^2}{k^2 \beta^2 \operatorname{Im} q_\varepsilon \operatorname{Im} q_\mu} > 0$$

$\square$

In the same manner we can show the following proposition for the case of general media. We state without proof:

**Proposition V.21 (General media).** *Given the wavenumber $k > 0$. Assume that*

$$\varepsilon, \mu \in \{re^{i\varphi} : r \in [r_1, r_2], \varphi \in [0, \varphi_2]\}$$

*almost everywhere with radii $1 < r_1 < r_2$ and angles $0 < \varphi_2 < \phi < \pi/4$. For a sufficiently small positive constant $\lambda$ such that $k^2\beta^2 < \lambda$ almost everywhere there exist positive constants $c_P, c_Q$ such that the contrasts satisfy*

$$\operatorname{Re}\left[e^{-i\phi}(\tilde{P}x) \cdot \bar{x}\right] \geq c_P|x|^2 \quad \text{and} \quad \operatorname{Re}\left[e^{-i\phi}(Qx) \cdot \bar{x}\right] \geq c_Q|x|^2$$

*and, furthermore,*

$$\operatorname{Im}\left[(\tilde{P}x) \cdot \bar{x}\right] \geq 0 \quad \text{and} \quad \operatorname{Im}\left[(Qx) \cdot \bar{x}\right] \geq 0.$$

## 5. Numerical Experiments

### 5.1. Visualization of the far field pattern

For reasons of simplicity we consider the achiral non–magnetic case in this subsection. In that case our model of the vector Helmholtz equation is redundant and it is sufficient to look at the scalar Helmholtz equation.

$$\Delta u + \kappa^2 u = 0.$$

Given a bounded scattering obstacle $D \subset \mathbb{R}^2$, $\kappa$ is defined such that $\kappa = k = \omega\sqrt{\varepsilon_0\mu_0}$ in the exterior of $\overline{\Omega}$. We illuminate $D$ by a plane wave. The far field pattern of the scattered field $u^s$ is given by

$$u^\infty(\hat{x}) = \frac{e^{i\pi/4}}{\sqrt{8\pi k}} \int_{\partial D} u^s(y)\frac{\partial e^{-ik\hat{x}\cdot y}}{\partial \nu(y)} - \frac{\partial u^s}{\partial \nu}(y)e^{-ik\hat{x}\cdot y}\,\mathrm{d}s(y). \qquad (5.28)$$

**Solving the direct problem**

In order to generate data we solve the direct problem numerically by finite elements. The computational domain is a circle with radius 2: $B := B(0,2)$. We use a variational formulation of the Helmholtz equation which we derive as follows. Given the free space wave number $k$ and a constrast $q$ with $\operatorname{supp} q \subset B$. For an incident field $u^i$ determine fields $u^s$ and $u$ satisfying

$$\begin{aligned}
\Delta u^i + k^2 u^i &= 0 & &\text{in } \overline{B}^c, \\
\Delta u^s + k^2 u^s &= 0 & &\text{in } \overline{B}^c, \text{ radiating}, \\
\Delta u + k^2(1+q)u &= 0 & &\text{in } B, \qquad (5.29)\\
\tfrac{\partial u^s}{\partial \nu} + \tfrac{\partial u^i}{\partial \nu} &= \tfrac{\partial u}{\partial \nu} & &\text{on } \partial B, \\
u^s + u^i &= u & &\text{on } \partial B.
\end{aligned}$$

Introduce the Dirichlet–Neumann operator $\Lambda\colon \lambda \to \frac{\partial v}{\partial \nu}$ where $v$ solves the exterior Dirichlet problem for the Helmholtz equation

$$\begin{aligned}
\Delta v + k^2 v &= 0 & &\text{in } \overline{B}^c, \text{ radiating}, \\
v &= \lambda & &\text{on } \partial B.
\end{aligned}$$

# 5. Numerical Experiments

In this case $\Lambda$ admits a series representation which can be used for a numerical implementation. We multiply equation (5.29) with a test function $\psi$ and integrate over $B$. Using integration by parts, the transmission conditions and the Dirichlet–Neumann operator we derive

$$\iint_B \nabla u \cdot \nabla \psi - k^2(1+q)u\psi \,\mathrm{d}x - \int_{\partial B} \Lambda u\,\psi \,\mathrm{d}s = \int_{\partial B} f^i \psi \,\mathrm{d}s \qquad (5.30)$$

with $f^i := \frac{\partial u^i}{\partial \nu} - \Lambda u^i$. We use standard finite elements to determine numerical solutions. The scattered field is obtained by subtracting the incident field $u^i$.

## Computation of the far field pattern

The scatterer is represented by the contrast $q$ and its support. We solve the direct problem for a finite set of plane waves $u_j^i(x) = e^{ik\,d_j \cdot x}$ with vectors $d_j = (\cos\theta_j, \sin\theta_j)^\top$ where $\theta_j = 2\pi j/n$, $j = 1\ldots n$. For every solution $u_j$ of the variational equation we evaluate the integral (5.28) for $u^s = u_j^s := u_j - u_j^i$ for unit vectors $\hat{x}_l = (\cos\phi_l, \sin\phi_l)^\top$, $\phi_l = 2\pi l/n$, $l = 1,\ldots,n$. We rather integrate over $\{|x| = 2\}$ than the boundary of the scatterer. The result is a matrix representing the far field pattern evaluated at the points $(\phi_l, \theta_k)$ for $k, l = 1,\ldots, n$.

## Reciprocity and symmetry

We denote by $u^\infty(\hat{x}, d)$ the far field pattern $u^\infty(\hat{x})$ induced by the incident field $u^i(x) = e^{ik\,d \cdot x}$ with direction of incidence $d = (\cos\theta, \sin\theta)^\top$ for an angle $\theta \in [0, 2\pi]$. The well known reciprocity relation holds:

$$u^\infty(\hat{x}, d) = u^\infty(-d, -\hat{x}), \qquad \hat{x}, d \in \mathbb{S}^1.$$

Here, $\hat{x} \in \mathbb{S}^1$ can be represented as $\hat{x} = (\cos\phi, \sin\phi)^\top$ with an angle $\phi \in [0, 2\pi]$. Then $-\hat{x} = (\cos(\phi + \pi), \sin(\phi + \pi))^\top$. We identify the unit vectors $\hat{x}$ and $d$ with $\phi$ and $\theta$, respectively, and denote by $u^\infty(\phi, \theta)$ the far field pattern $u^\infty(\hat{x}, d)$. Then the reciprocity relation reads

$$u^\infty(\phi, \theta) = u^\infty(\theta + \pi, \phi + \pi), \qquad \phi, \theta \in [0, 2\pi]. \qquad (5.31)$$

We interpret the far field pattern as a $(2\pi, 2\pi)$–periodic function on $\mathbb{R}^2$. The geometric meaning of the reciprocity relations is a symmetry with respect to the line $\{\theta = \phi + \pi\}$. We verifiy this by shifting the far field pattern and define:

$$\tilde{u}^\infty(\phi, \theta) := u^\infty(\phi, \theta + \pi) \quad \text{for } \phi, \theta \in [-\pi, \pi].$$

Now we can show that $\tilde{u}^\infty$ is symmetric; that is, $\tilde{u}^\infty(\phi, \theta) = \tilde{u}^\infty(\theta, \phi)$:

$$\tilde{u}^\infty(\phi, \theta) = u^\infty(\phi, \theta+\pi) \stackrel{(5.31)}{=} u^\infty(\theta+2\pi, \phi+\pi) = u^\infty(\theta, \phi+\pi) = \tilde{u}^\infty(\theta, \phi).$$

This proves the symmetry property of $u^\infty$.

**Rotation**

What happens to $u^\infty$ if the scatterer is rotated around the origin by an angle $\varphi \in [0, 2\pi]$. An observation which was made at a point $(\phi, \theta)$ can be seen at the point $(\phi + \varphi, \theta + \varphi)$ after the rotation. Obviously $u^\infty$ is shifted by $\sqrt{2}\varphi$ along the line $\{\phi = \theta\}$, more precisely:

$$u_\varphi^\infty(\phi, \theta) = u_0^\infty(\phi - \varphi, \theta - \varphi) \tag{5.32}$$

where $u_0^\infty$ and $u_\varphi^\infty$ are the far field patterns of the obstacle in the original and rotated position, respectively. Figure V.3 shows the shifting of the far field pattern of a kite, which is rotated.

From (5.32), we immediately see the well known property of sperical centered scattering obstacles: From every point they look the same. In this case $u_\varphi^\infty = u_0^\infty$ for every $\varphi \in [0, 2\pi]$. Hence, the far field pattern consists of lines parallel to $\{\phi = \theta\}$.

Furthermore, for a regular centered polygon with $n$ edges we have $u_\varphi^\infty = u_0^\infty$ for $\varphi = 2\pi k/n$, $k = 0, \ldots, n$. Hence, we can identify a certain pattern, which is repeated $n$ times within the square $[0, 2\pi] \times [0, 2\pi]$. Figure V.4 shows the far field patterns of a rectangle, a hexagon, an octogon and a circle.

**Mirroring**

Given an object which is symmetric with respect to $\{x_2 = 0\}$. The corresponding far field pattern doesn't change after mirroring. We can imagine the far field pattern as a result of measurements which are made as follows. An emitting and recieving device are moved on a circle around

5. Numerical Experiments 133

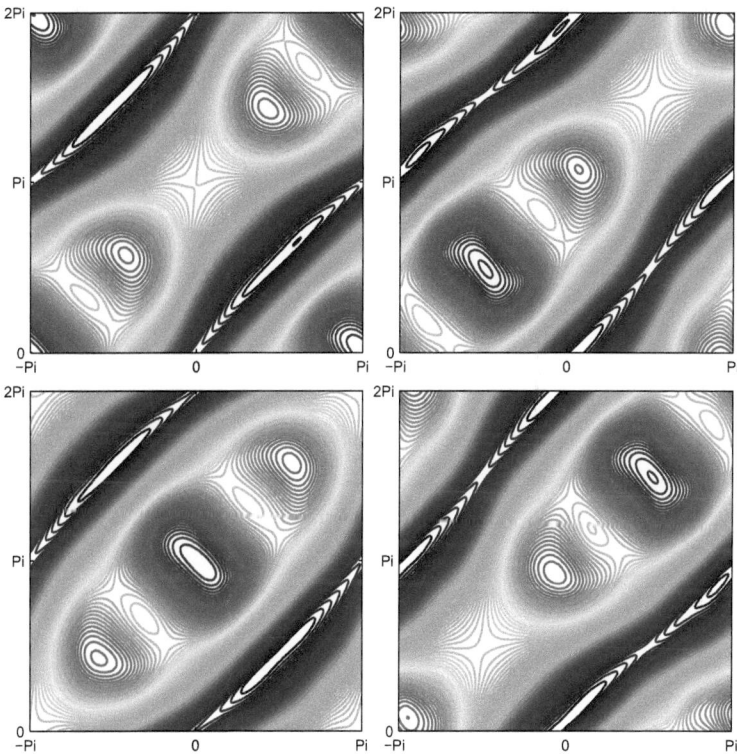

Figure V.3: Re $(u^\infty)$ for a kite rotated by $0, \pi/2, \pi$ and $3\pi/2$.

134                    Factorization Method for the vector Helmholtz case

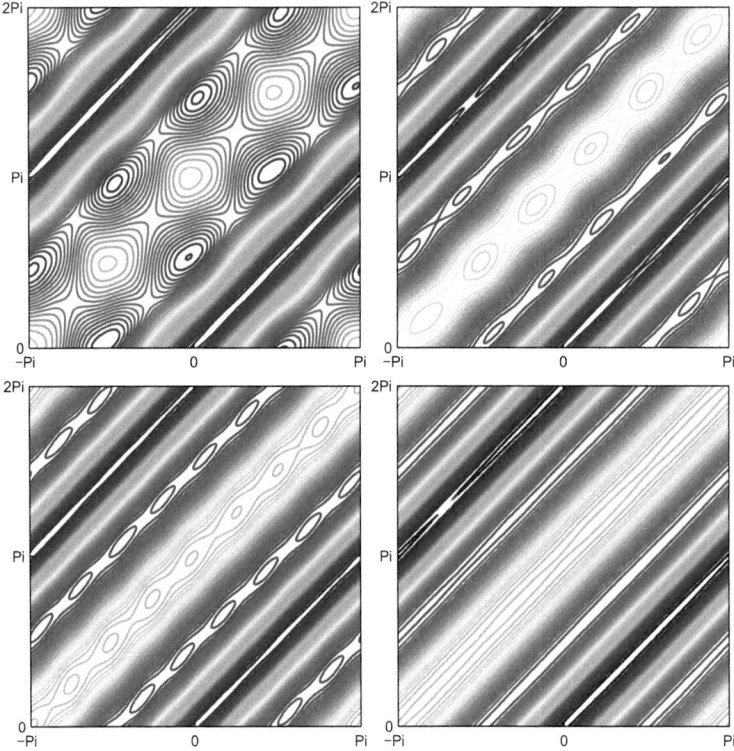

Figure V.4: Re $(u^\infty)$ for centered polygons with 4, 6 and 8 vertices and a circle.

## 5. Numerical Experiments

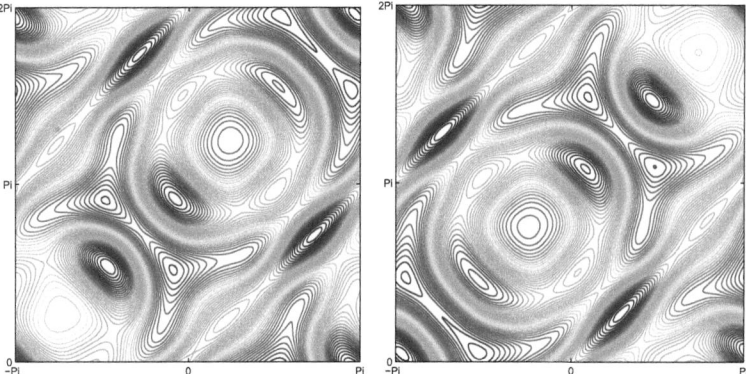

Figure V.5: Re$(u^\infty)$ for a kite and a rectangle mirrored at $x_2 = 0$.

the obstacle. In general, the sense of rotation – positive or negative – has an influence on the measurements. But in the symmetric case the measurements are the same wether they are taken by positive or negative rotation. This means: $u^\infty(\phi, \theta) = u^\infty(2\pi - \phi, 2\pi - \theta)$ for all $\phi, \theta \in [0, 2\pi]$. Geometrically we can rotate the far field pattern by $\pi$ and it is congruent to the original one. The plots in the left column of Figure V.3 show this symmetry.

As consequence, for general obstacles the far field pattern $u_m^\infty$ of the mirrored obstacle is given by

$$u_m^\infty(\phi, \theta) = u^\infty(2\pi - \phi, 2\pi - \theta) \quad \text{for } \phi, \theta \in [0, 2\pi].$$

Figure V.5 shows the far field patterns for a kite and a rectangle, mirrored at $\{x_2 = 0\}$.

### Translation

In this paragraph we study the transformation of $u^\infty$ under translation. Therefore, let $D$ be the scattering obstacle and $D_\tau := D + \tau$ the shifted obstacle with translation vector $\tau \in \mathbb{R}^2$. We illuminate $D$ with a plane wave $u^i(x) = e^{ik\,d\cdot x}$ with direction of incidence $d$. This causes a scattered field $u^s(\cdot, d)$ and we can compute the far field pattern $u^\infty$ by

$$u^\infty(\hat{x}, d) = \int_{\partial D} u^s(y,d) \frac{\partial e^{-ik\,\hat{x}\cdot y}}{\partial \nu(y)} - \frac{\partial u^s(y,d)}{\partial \nu(y)} e^{ik\,\hat{x}\cdot y} \, ds(y).$$

Now, we introduce the shifted incident field

$$u^i_\tau(x) = u^i(x - \tau) = u^i(x)e^{-ik\,\tau \cdot d}$$

and illuminate $D_\tau$ by $u^i_\tau$. The induced scattered field $u^s_\tau(\cdot, d)$ is the shifted version of $u^s$ and satisfies

$$u^s_\tau(x, d) = u^s(x - \tau, d).$$

For the computation of the far field pattern caused by $D_\tau$ we have to use the incident field $u^i = u^i_\tau e^{ik\,\tau \cdot d}$. By linearity, the scattered field $v^s(\cdot, d)$ caused by $u^i$ is given by

$$v^s(x, d) = e^{ik\,\tau \cdot d} u^s_\tau(x, d) = e^{ik\,\tau \cdot d} u^s(x - \tau, d).$$

Now, we can compute the far field pattern $v^\infty$:

$$v^\infty(\hat{x}, d) = \int_{\partial D_\tau} v^s(y,d) \frac{\partial e^{-ik\,\hat{x}\cdot y}}{\partial \nu(y)} - \frac{\partial v^s(y,d)}{\partial \nu(y)} e^{-ik\,\hat{x}\cdot y} \, ds(y)$$

$$= e^{ik\tau \cdot d} \int_{\partial D_\tau} u^s(y - \tau, d) \frac{\partial e^{-ik\,\hat{x}\cdot y}}{\partial \nu(y)} - \frac{\partial u^s(y - \tau, d)}{\partial \nu(y)} e^{ik\,\hat{x}\cdot y} \, ds(y)$$

$$= e^{ik\,\tau \cdot d} \int_{\partial D} u^s(y, d) \frac{\partial e^{-ik\,\hat{x}\cdot (y+\tau)}}{\partial \nu(y)} - \frac{\partial u^s(y,d)}{\partial \nu(y)} e^{-ik\,\hat{x}\cdot(y+\tau)} \, ds(y)$$

$$= e^{ik\,(d-\hat{x})\cdot \tau} u^\infty(\hat{x}, d).$$

This formula shows that the shifted obstacle can be distinguished from the original one by the phase shift of the far field patterns. Figure V.7 shows – up to numerical inaccuracy: $|v^\infty| = |u^\infty|$. Figure V.6 shows the transformation of the far field pattern when a centered circle is moved into the direction $(1,0)^\top$.

## 5.2. Reconstruction of the scatterer

In this case we solve the full vector Helmholtz equation with the material parameters $\varepsilon$, $\mu$ and $\beta$. The variational equation is a more general vector

5. Numerical Experiments 137

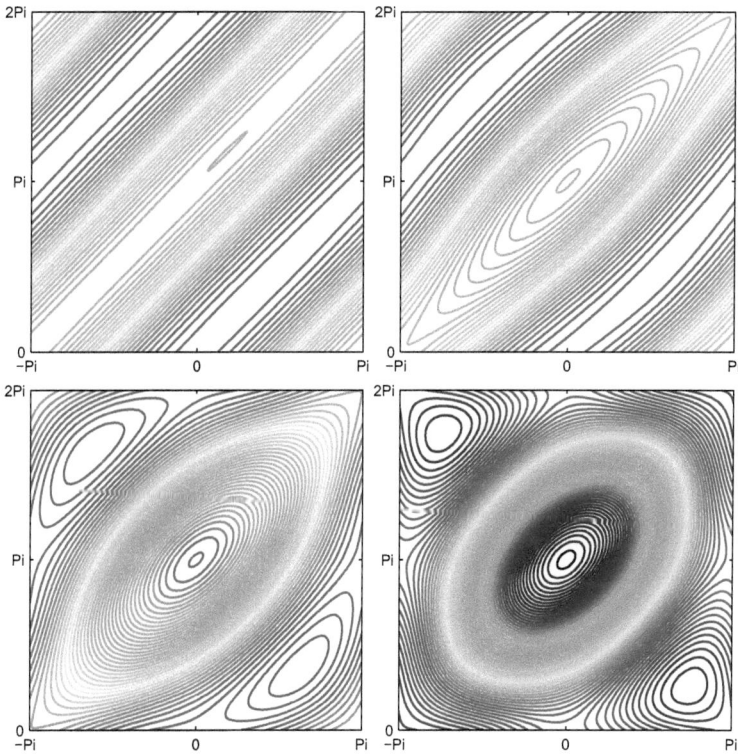

Figure V.6: $\arg(u^\infty)$ for a centered circle shifted to the right.

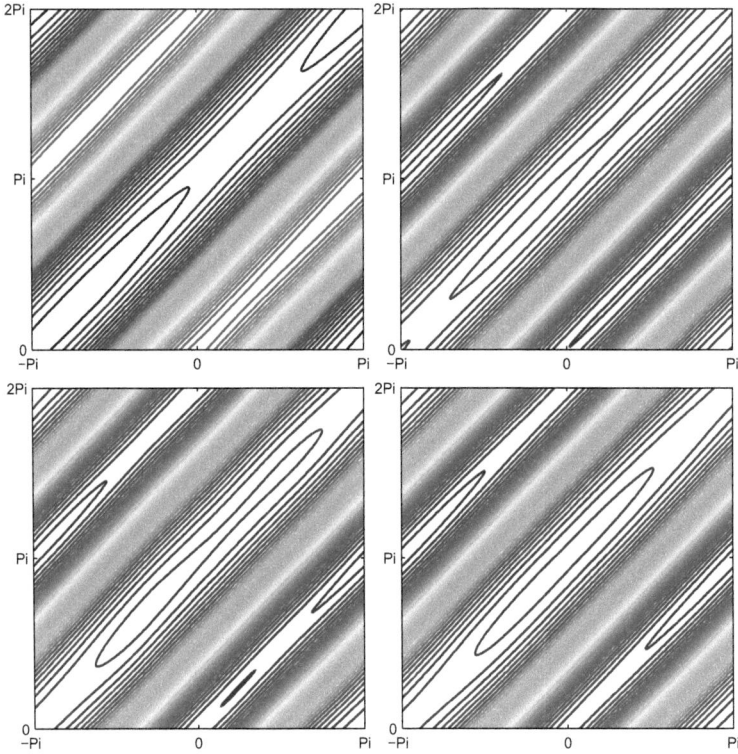

Figure V.7: $|u^\infty|$ for a centered circle shifted to the right.

## 5. Numerical Experiments

version of (5.30). With the contrasts $P$ and $Q$ the variational formulation reads

$$\iint_B \left((I-P)\nabla u\right)\cdot\psi - k^2\left((I+Q)u\right)\cdot\psi\,\mathrm{d}x - \int_{\partial B}(\Lambda u)\cdot\psi\,\mathrm{d}s = \int_{\partial B} f^i\cdot\psi\,\mathrm{d}s$$

where $f^i := \frac{\partial u^i}{\partial \nu} - \Lambda u^i$. Here, $u, u^i$ and $\psi$ are 2D–vectors, $P$ and $Q$ are $(4\times 4)$– and $(2\times 2)$–matrices, respectively, whose support is contained in $B(0,3/2)$. The gradient, the normal derivative and the Dirichlet–Neumann operator are applied componentwise.

In order to compute the far field operator we need to solve the direct problem for Herglotz wave functions of the form

$$u^i(x) = \int_{\mathbb{S}^1} p(d)e^{ik\,d\cdot x}\,\mathrm{d}s(d)\,,\qquad x\in\mathbb{R}^2,$$

with density $p \in L^2(\mathbb{S}^1,\mathbb{C}^2)$. For fixed $n \in \mathbb{N}$ we use a $(4n+2)$-dimensional subspace of $L^2(\mathbb{S}^1,\mathbb{C}^2)$ by choosing the basis functions

$$\mathcal{B} := \left\{ b e^{im\theta} \,:\, b \in \left\{\begin{pmatrix}1\\0\end{pmatrix},\begin{pmatrix}0\\1\end{pmatrix}\right\}, m=-n,\ldots,n, \theta\in[0,2\pi]\right\}.$$

We solve the direct problem for a finite set of Herglotz wave functions $u^i_p$ defined by

$$u^i_p(x) = \int_{\mathbb{S}^1} p(d)e^{ik\,d\cdot x}\,\mathrm{d}s(d)\,,\qquad p\in\mathcal{B}.$$

We evaluate the integral (5.28) on the circle $\partial B(0,3/2)$ to compute the far field pattern $u_p^\infty$. The scattering obstacle(s) is(are) situated somewhere inside this circle. Furthermore, we compute the Fourier coefficients $\alpha_q^p$ in the representation

$$u_p^\infty(\hat{x}) = \sum_{q\in\mathcal{B}} \alpha_q^p q(\hat{x})\,,\qquad \hat{x}\in\mathbb{S}^1.$$

The resulting matrix $\mathcal{F}_n := (\alpha_q^p)_{q,p\in\mathcal{B}}$ is an approximation of the far field operator $\mathcal{F}$. We determine an eigensystem $\{(\lambda_l,\psi_l), l=1,\ldots 4n+2\}$ of $|\mathrm{Re}\,\mathcal{F}_n| + \mathrm{Im}\,\mathcal{F}_n$ and implement the test function $\phi_z$. Finally, we evaluate the function

$$W(z) = \left[\sum_{j=1}^{4n+2} \frac{|(\phi_z,\psi_j)|^2}{\lambda_j}\right]^{-1}$$

140  Factorization Method for the vector Helmholtz case

| shape | | $\varepsilon$ | $\mu$ | $\beta$ |
|---|---|---|---|---|
| $R_1$ | rectangle | 1 | $0.01 + i$ | 0.01 |
| $K_1$ | kite | 0.1 | 1 | 0 |
| $K_2$ | kite | 0.1 | $1 + i$ | 0 |

Table V.2: Scatterers used for the experiments.

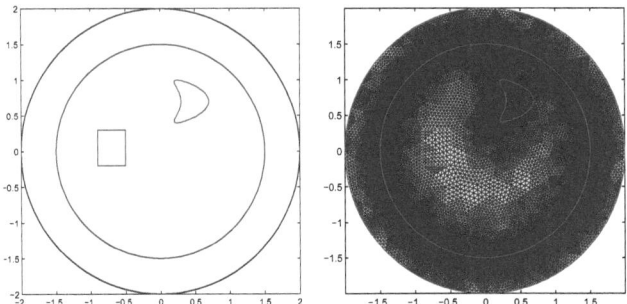

Figure V.8: Setting and grid for numerical experiments. $B(0,2)$ computational domain, $\partial B(0, 3/2)$ computation of far field pattern, scatterers: rectangle and kite.

on a mesh of points $z$ in $B$. Figure V.8 shows the setting and the grid for our numerical experiments. Two scatterers – a rectangle and a kite – are considered. In Table V.2 we describe the scatterers and their material parameters used for the experiments. The computational domain is $B(0,2)$ and the far field pattern is computed on $\partial B(0, 3/2)$. Furthermore we use different eigensystems (ES): ES1 refers to the eigensystem of $\mathcal{F}_\#$ and ES2 to the eigensystem of $\operatorname{Im} \mathcal{F}$. In all reconstruction we use a $(102 \times 102)$-matrix to approximate $\mathcal{F}$.

In the following we show some results. We start with the rectangle $R_1$ which admits chirality and a high (and complex valued) contrast in $\mu$. Figure V.9 shows a good reconstruction with the eigensystem of $\operatorname{Im} \mathcal{F}$ which works also with little noise. The localization of the scatterer improves for larger values of the wave number $k$. (Figure V.10). Figure V.11 shows that the reconstruction is more precise when using the eigensystem of $\operatorname{Im} \mathcal{F}$ (ES2). Furthermore, with 5% noise the obstacle can still be located

## 5. Numerical Experiments

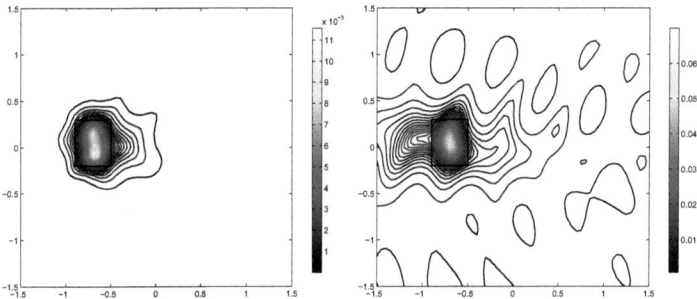

Figure V.9: $R_1$, ES2 without noise (left) and 1% noise (right).

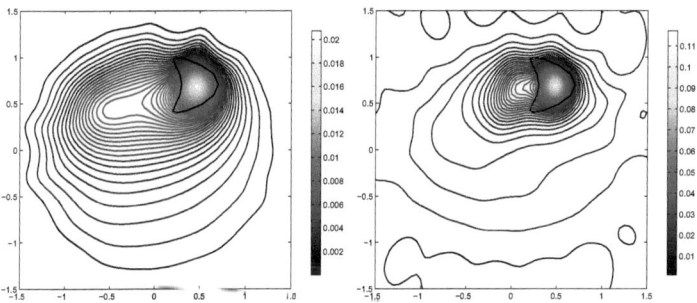

Figure V.10: $K_1$, ES1 for wave number $k = 2$ (left) and $k = 6$ (right).

quite well. Finally, we show what our implementation computes without scatterer. Figure V.12 shows the magnitudes which are in the range of $10^{-3}$ and $10^{-5}$, respectively. We expect that this is due to approximation error of the numerical scheme.

# Factorization Method for the vector Helmholtz case

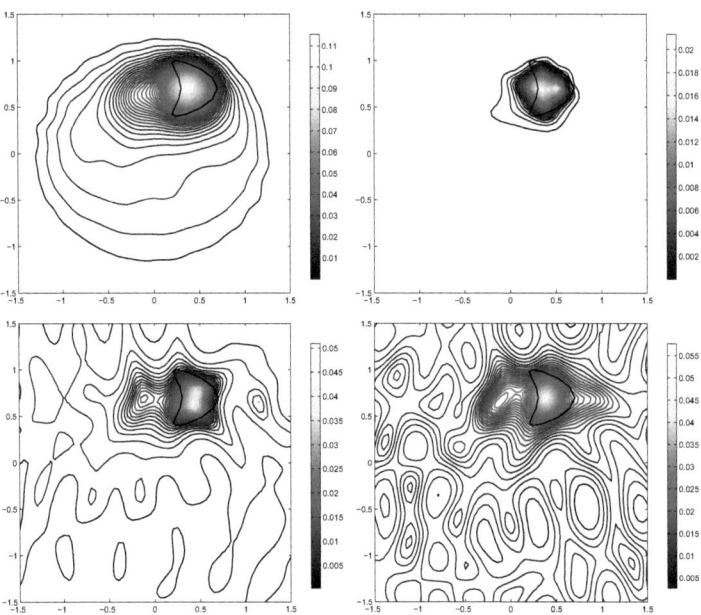

Figure V.11: $K_2$, ES1 without noise (upper left), $K_2$, ES2 without (upper right) and with noise: 1% (lower left), 5% (lower right).

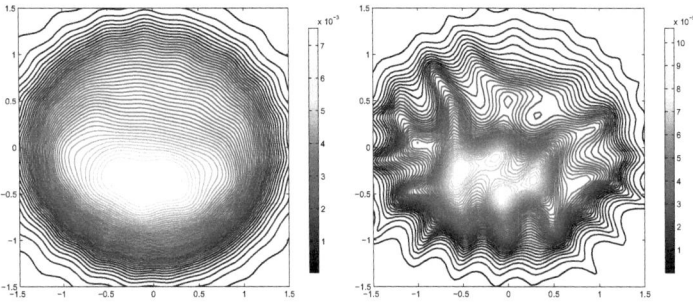

Figure V.12: Inverse scattering without scatterer: ES1 (left), ES2 (right).

CHAPTER VI

# Outlook: Periodic chiral media

This chapter forms – together with following – the final part of this work. Without going into details and without proofs we like to show, how to apply the results for scattering from bounded obstacles to the Factorization method for periodic chiral media. We start with the Factorization method for non–magnetic achiral periodic media.

We consider the scattering of electromagnetic waves from a biperiodic chiral structure $\Omega' \subset \mathbb{R}^3$, which is periodic in the $x_1$- and $x_2$-direction and has finite extension in the $x_3$-direction. Without loss if generality we assume that $\Omega'$ is $\Lambda$-periodic with $\Lambda = (2\pi, 2\pi, 0)^\top$. The material parameters $\varepsilon, \mu$ and $\beta$ are also $\Lambda$-periodic and $\varepsilon \equiv \mu \equiv 1$ in $\mathbb{R}^3 \setminus \overline{\Omega'}$ and $\beta \equiv 0$ in $\mathbb{R}^3 \setminus \overline{\Omega'}$. We introduce the unit cell $D := (-\pi, \pi)^2 \times \mathbb{R}$ and $\Omega := \Omega' \cap D$. An incident field irradiates $\Omega'$. In this setting we are interested in quasi–periodic solutions to the scattering problem. The inverse problem is to reconstruct $\Omega$ from near field measurements. Figure VI.1 shows the setting.

Due to the periodicity it is sufficient to solve the problem in the unit cell $D$. Sandfort [37] applies the Factorization method to this problem for achiral non–magnetic materials. He shows that it is possible to apply the methods and techniques used for the case of scattering by bounded obstacles. We can consider his results as a kind of generalization – from bounded obstacles to biperiodic structures. Combining his results and the generalization to chiral media found in chapters II and III yields the Factorization method for chiral biperiodic structures.

In the sequel we briefly state the main equations and operators for the

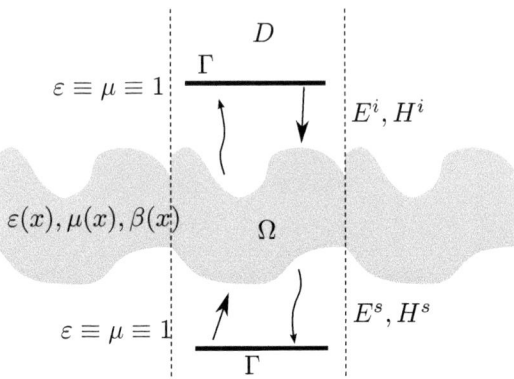

Figure VI.1: Setting for the scattering from peridioc structures.

periodic chiral case. We don't use the same notation but refer to the PhD thesis [37] for all technical details concerning for example regularity assumptions for $\Omega'$ and $\Omega$, the Sobolev spaces for quasi–periodic functions with $L^2$–curl, traces, Green's formula, the quasi–periodic Green's tensor for Maxwell's equations, the Rayleigh expansion (as radiation condition for outgoing waves), ...

**Weak formulation of the transmission problem**

Let $k > 0$. Given $g, h \in L^2(\Omega, \mathbb{C}^3)$ determine $\alpha$-quasi-periodic, radiating solution $u_\alpha \in H_{\alpha,\mathrm{loc}}(\mathrm{curl}, D)$ such that

$$\iint_D \left[\left(\tfrac{1}{\varepsilon} - k^2 \mu \beta^2\right) \mathrm{curl}\, u_\alpha - k^2 \mu \beta\, u_\alpha\right] \cdot \mathrm{curl}\, \psi_{-\alpha}\, \mathrm{d}x$$

$$- k^2 \iint_D \left[\mu \beta\, \mathrm{curl}\, u_\alpha + \mu u_\alpha\right] \cdot \psi_{-\alpha}\, \mathrm{d}x = \iint_\Omega g \cdot \psi_{-\alpha} + h \cdot \mathrm{curl}\, \psi_{-\alpha}\, \mathrm{d}x$$

for all $(-\alpha)$-quasi-periodic test functions $\psi_{-\alpha} \in H_{-\alpha,\mathrm{c}}(\mathrm{curl}, D)$ (with compact support).

**Existence and Uniqueness**

Sandfort uses the $\alpha$-quasi-periodic Green's tensor $G_{\alpha,k}$ for the Maxwell operator $\mathrm{curl}^2 - k^2 \mathrm{id}$ to define the volume potential: For $h \in L^2(\Omega, \mathbb{C}^3)$

define $w_\alpha \in H_{\alpha,\text{loc}}(\text{curl}, D)$ by

$$w_\alpha := \text{curl} \iint_\Omega G_{-\alpha,k}(x,y) h(y) \, \mathrm{d}(y)$$

which is radiating and solves

$$\iint_D \text{curl}\, w_\alpha \cdot \text{curl}\, \psi_{-\alpha} - k^2 w_\alpha \cdot \psi_{-\alpha} \, \mathrm{d}x = \iint_\Omega g \cdot \text{curl}\, \psi_{-\alpha} \, \mathrm{d}x$$

for all $\psi_{-\alpha} \in H_{-\alpha,\text{c}}(\text{curl}, D)$. As in the case of bounded scattering we need a second volume potential: For $g \in L^2(\Omega, \mathbb{C}^3)$ define $v_\alpha \in H_{\alpha,\text{loc}}(\text{curl}, D)$ by

$$v_\alpha := \iint_\Omega G_{-\alpha,k}(x,y) g(y) \, \mathrm{d}y$$

which is radiating and solves

$$\iint_D \text{curl}\, w_\alpha \cdot \text{curl}\, \psi_{-\alpha} - k^2 w_\alpha \cdot \psi_{-\alpha} \, \mathrm{d}x = \iint_\Omega g \cdot \psi_{-\alpha} \, \mathrm{d}x$$

With these volume potentials one can derive an integro–differential equation, reformulate it with appropriate operators and use the Fredholm theory. At least for absorbing media all theorems and proofs are analogous and will end up with the following

**Theorem VI.1.** *Assume that*

*(a) $\varepsilon, \mu, \beta \in L^\infty(D)$, $\text{Im}\, \beta = 0$ s.t. $\varepsilon \equiv \mu \equiv 1$, $\beta \equiv 0$ in $D \setminus \overline{\Omega}$,*

*(b) $\text{Re}\,\mu \geq c_1$, $\text{Re}\,(1/\varepsilon) \geq c_2$, $k^2 \beta^2 \frac{|\varepsilon|^2 |\mu|^2}{\text{Re}\,\varepsilon \, \text{Re}\,\mu} \leq c_3 < 1$ a.e.,*

*(c) $\text{Im}\,\varepsilon > 0$ and $\text{Im}\,\mu \geq 0$ a.e. in $\Omega$.*

*For every $(g,h) \in L^2(\Omega, \mathbb{C}^3)^2$ there exists a unique radiating solution $v_\alpha \in H_{\alpha,\text{loc}}(\text{curl}, D)$ of the weak transmission problem. For any compact set $B$ with $D \supset B \supset \overline{\Omega}$ there exists $C > 0$ such that*

$$\|v_\alpha\|_{H_\alpha(\text{curl},B)} \leq C \|(g,h)\|_{L^2(\Omega,\mathbb{C}^3)^2} \quad \text{f.a. } (g,h) \in L^2(\Omega, \mathbb{C}^3)^2.$$

For non–absorbing bounded obstacles we used the unique continuation principle. This cannot be done in the periodic case. Ammari and Bao [3] show existence and uniqueness for the real–valued case for all but possibly a discrete set of frequencies.

## Reconstruction of the scatterer

We consider $\alpha$-quasi-periodic incident fields which originate from magnetic dipoles on the surface $\Gamma$. For a vector moment function $\varphi \in L^2(\Gamma, \mathbb{C}^3)$ define
$$u_\alpha^i(x) = \int_\Gamma \overline{G_{-\alpha,k}(y,x)} \varphi(y)\, \mathrm{d}s(y)\,, \quad x \in D \setminus \Gamma.$$
$u_\alpha^i$ generates the scattered field
$$u_\alpha(x) = \int_\Gamma u_\alpha^s(x,y) \varphi(y)\, \mathrm{d}s(y)\,, \quad x \in D,$$
where $u_\alpha^s(\cdot, y)$ is the scattering response of a magnetic dipole at $y \in \Gamma$.

**Inverse problem** Given the wave number $k > 0$ and all scattered fields $u_\alpha$ on $\Gamma$ for all moment functions $\varphi \in L^2(\Gamma, \mathbb{C}^3)$ determine $\Omega$.

The treatment of the inverse problem for absorbing media is analogous to chapter 3. Define the near field operator $\mathcal{N} \colon L^2(\Gamma, \mathbb{C}^3) \to L^2(\Gamma, \mathbb{C}^3)$ by
$$(\mathcal{N}\varphi)(x) := \int_\Gamma u_\alpha^s(y,x) \varphi(y)\, \mathrm{d}s(y)\,, \quad x \in \Gamma,$$
and show the factoriation
$$\mathcal{N} = \mathcal{H}^* \mathcal{T} \mathcal{H}.$$
Here $\mathcal{H} \colon L^2(\Gamma, \mathbb{C}^3) \to L^2(\Omega, \mathbb{C}^3)^2$, $\mathcal{H}\varphi = (\mathcal{H}_1\varphi, \mathcal{H}_2\varphi)^\top$ with
$$(\mathcal{H}_1\varphi)(x) := \int_\Gamma \overline{G_{-\alpha,k}(y,x)} \varphi(y)\, \mathrm{d}s(y)\,, \quad \mathcal{H}_2\varphi = \mathrm{curl}\, \mathcal{H}_1\varphi$$
for $x \in \Gamma$ and the adjoint operator $\mathcal{H}^* \colon L^2(\Omega, \mathbb{C}^3)^2 \to L^2(\Gamma, \mathbb{C}^3)$ is given by $\mathcal{H}^* g = \mathcal{H}_1^* g_1 + \mathcal{H}_2^* g_2$ with
$$(\mathcal{H}_1^* g_1)(x) = \iint_\Omega G_{\alpha,k}(x,y) g_1(y)\, \mathrm{d}y\,, \quad x \in \Gamma,$$
and
$$(\mathcal{H}_2^* g_2)(x) = \mathrm{curl} \iint_\Omega G_{\alpha,k}(x,y) g_2(y)\, \mathrm{d}y\,, \quad x \in \Gamma.$$
$H^* g = w_\alpha|_\Gamma$ where $w_\alpha$ is weak radiating solution of
$$\mathrm{curl}^2 w_\alpha - k^2 w_\alpha = g_1 + \mathrm{curl}\, g_2.$$

Furthermore, $\mathcal{H}^*$ characterizes $\Omega$: For $z \in D$ and fixed $p \in \mathbb{C}^3 \smallsetminus \{0\}$ define $\phi_z(x) = k^2 G_\alpha(z,y)p$. Then $z \in \Omega$ if, and only if $\phi_z \in \mathcal{R}(\mathcal{H}^*)$. Finally, the middle operator $\mathcal{T}: L^2(\Omega, \mathbb{C}^3) \to L^2(\Omega, \mathbb{C}^3)^2$ is given by

$$\mathcal{T}f := \begin{pmatrix} k^2 q_\mu & k^2 \mu\beta \\ k^2 \mu\beta & q_\varepsilon + k^2 \mu\beta^2 \end{pmatrix} \begin{pmatrix} f_1 + v_\alpha \\ f_2 + \operatorname{curl} v_\alpha \end{pmatrix}$$

with the constrast $q_\varepsilon = 1 - \varepsilon^{-1}$ and $q_\mu = \mu - 1$. $v_\alpha$ is the $\alpha$-quasi-periodic, radiating weak solution of

$$\iint_D \left[ \left(\tfrac{1}{\varepsilon} - k^2 \mu\beta^2\right) \operatorname{curl} v_\alpha - k^2 \mu\beta v_\alpha \right] \cdot \operatorname{curl} \psi_{-\alpha}\, dx$$

$$- k^2 \iint_D \left[ \mu\beta \operatorname{curl} v_\alpha + \mu v_\alpha \right] \cdot \psi_{-\alpha}\, dx$$

$$= \iint_\Omega k^2 [q_\mu f_1 + \mu\beta f_2] \cdot \psi_{-\alpha} + \left[ (q_\varepsilon + k^2 \mu\beta^2) f_2 + k^2 \mu\beta f_1 \right] \cdot \operatorname{curl} \psi_{-\alpha}\, dx$$

for all $\psi_{-\alpha} \in H_{-\alpha,c}(\operatorname{curl}, D)$ with compact support. The proof of the properties of $\mathcal{T}$ are analogous to the case of bounded obstacles and one can finally show

**Theorem VI.2 (Absorbing Media).** *Assume that there exist constants* $c_\varepsilon, c_\mu > 0$ *such that*

$$\operatorname{Im} q_\mu \geq c_\mu \quad \text{and} \quad \operatorname{Im} q_\varepsilon \geq c_\varepsilon \quad \text{a.e. in } \Omega.$$

*Then*

*(a)* $\operatorname{Im}(\mathcal{T}f, f) \geq 0$ *for* $f \in L^2(\Omega, \mathbb{C}^3)^2$.

*(b) There exists* $c > 0$ *such that* $\operatorname{Im}(\mathcal{T}f, f) \geq c\|f\|^2$ *for* $f \in \mathcal{R}(\mathcal{H})$.

*(c) The ranges of* $(\operatorname{Im}\mathcal{N})^{1/2}$ *and* $\mathcal{H}^*$ *coincide.*

For a point $z \in D$ we conlude:

$$z \in \Omega \iff \phi_z \in \mathcal{R}\left((\operatorname{Im}\mathcal{N})^{1/2}\right)$$

where $\phi_z$ is defined as above.

CHAPTER VII

# Conclusions

This work deals with several aspects of inverse scattering for chiral materials. Scattering from a bounded obstacle is studied in detail: both the direct and the inverse problem. The special case of scattering from a homogeneous chiral sphere is done analytically. Scattering by chiral cylinder is used to motivate the factorization method for the vector Helmholtz equation. Numerical examples serve as proof of concept and illustrate the theoretical results. Finally, scattering from periodic chiral structures is another possible application of the generalized factorization method.

In the literature many results of existence and uniqueness for the direct transmission problem or related problems can be found. Nevertheless, we generalized the method Kirsch proposed and adapted the integro differential equation approach. The key to success was to allow only real valued chirality parameter $\beta$. Then the volume potential solutions could be modified appropriatly. The assumptions on the parameters $\varepsilon, \mu$ and $\beta$ turned out to coincide with those for the achiral case. Additional assumptions were always of the form: $k^2\beta^2$ sufficiently small, which is common in the chiral literature.

We generalized the Factorization method for the treatment of the inverse problem. Therefore we introduced the Herglotz operator with two components and in consequence the middle operator could be written as matrix–vector–multiplication. In the case of absorbing media, with the technique of completing the square we could show important properties of the middle operator $\mathcal{T}$ in a quite straight forward manner. The real valued case is more complicated since $H(\mathrm{curl}, \Omega)$ is not compactly embedded in $L^2(\Omega, \mathbb{C}^3)$. Therefore additional smoothness assumptions on the parameters are requested and a Helmholtz decomposition is used.

For the scattering by a homogeneous chiral sphere we could exploit two concepts. Firstly, Bohren's decompostion into Beltrami fields $-Q_R = E + iH$ and $Q_L = E - iH$ with wave numbers $\kappa_L$ and $\kappa_R$ - and secondly, series expansions. We were able to compute explicitely the scattered field and the far field pattern. Furthermore, we computed the eigensystem of the far field operator explicitly. The achiral eigenfunctions are the vector spherical harmonics $U_n^m$ and $V_n^m$. The chiral eigenfunctions are the linear combinations $U_n^m + iV_n^m$ and $U_n^m - iV_n^m$. The corresponding eigenvalues depend on $\kappa_L$ and $\kappa_R$, respectively.

Scattering by an infinite chiral cylinder leads to the vector Helmholtz equation for $E_3$ and $H_3$ with perturbations in the div $\nabla$–term and the $k^2$–term. Here we applied the Factorization method by combining results for the Helmholtz equation with only one perturbation. This chapter can be seen as application, since the methods and arguments in the proofs were already presented in chapters II and III. Nevertheless, the compact embedding of $H^1(\Omega)$ into $L^2(\Omega)$ made the proofs more simple. With standard finite elements we could generate far field measurements for testing an implementation of the Factorization method. The implementation is quite simple and the reconstructions show that the method works. The results depend on the contrasts, the wave number and the noise. Unfortunately, we could only work with our simulated measurements. Additionally, plots of far field patterns complemented the theory and illustrated what kind of information the far field pattern contains: the reciprocity relation causes symmetry and certain properties of the scatterer can be explained by symmetry arguments.

Finally, we are convinced that the generalization from achiral to chiral media can be done analogously for biperiodic structures. We formulated these ideas as outlook and gave the main steps without proof.

Future work includes the question how to weaken the assumptions for non–absorbing materials and numerical experiments with real data and numerical schemes for the 3D Maxwell's equations. Other types of obstacles could be of interest, for example objects with chiral coatings. A rigorous proof of the Factorization method for chiral periodic structures would be a good starting point to investigate scattering from chiral metamaterials. In contrast to the reconstruction of the scatterer, it would be interesting how the additional degree of freedom $\beta$ influences the possibilities for cloaking.

# List of Symbols

$\beta$ .......... chirality, page 11
$\mathcal{C}_1, \mathcal{C}_2$ ....... boundary integral operators, page 95
$\mathcal{F}$ .......... far field operator, page 44
$\mathcal{F}^*$ .......... adjoint far field operator, page 47
$\Gamma$ .......... boundary of $\Omega$, page 13
$\hat{x}$ .......... $x/|x| \in \mathbb{S}^2$, page 19
$\hat{\psi}$ .......... $(\psi, \operatorname{curl} \psi)^\top$ for functions $\psi$ with well defined curl, page 65
$H_\mathrm{c}(\operatorname{curl}, \mathbb{R}^3)$ . vector fields with weak curl and compact support, page 18
$H(\operatorname{curl}, \Omega)$ .. vector fields with weak curl, page 18
$\mathcal{H}, \mathcal{H}_1, \mathcal{H}_2$ ... Herglotz operators, page 52
$\mathcal{H}^*, \mathcal{H}_1^*, \mathcal{H}_2^*$ .. adjoint Herglotz operators, page 53
$\mathcal{H}^\dagger$ .......... $L^2_Q$-adjoint operator, page 58
$H_\mathrm{loc}(\operatorname{curl}, \mathbb{R}^3)$ vector fields with locally existing weak curl, page 18
$\kappa$ .......... generalized wave number, page 21
$\Lambda$ .......... Dirichlet–Neumann operator, page 130
$\mathcal{B}$ .......... magnetic induction, page 10
$\mathcal{D}$ .......... electric induction, page 10
$\mathcal{E}$ .......... electric field, page 10
$\mathcal{H}$ .......... magnetic field, page 10
$\mathcal{T}$ .......... middle operator, page 56
$\mu$ .......... relative permeability, page 11
$\mu_0$ .......... magnetic permeability in vacuum, page 10
$\|\cdot\|_\beta$ ......... equivalent norm on $H(\operatorname{curl}, \mathbb{R}^3)$, page 33
$\|\cdot\|_\beta$ ......... equivalent norm on $H(\operatorname{curl}, \mathbb{R}^3)$, page 67
$\|\cdot\|_{\mu\beta}$ ........ equivalent norm on $H(\operatorname{curl}, \mathbb{R}^3)$, page 73

$\nu$ ............ unit normal vector on $\Gamma$, page 14
$\mathcal{N}(A)$ ....... null space of an operator $A$, page 59
$\Omega$ ............ chiral body (bounded domain), page 13
$\omega$ ............ frequency, page 2
$\phi_z$ ........... test function for the reconstruction, page 75
$\mathcal{P}$ ............ orthongal projector on $X$, page 58
$\Pi$ ............ set of complex numbers: $\kappa \in \Pi$, page 21
$\mathcal{R}(A)$ ....... range of an operator $A$, page 59
$\mathcal{S}$ ............ scattering operator, page 48
sgn ........ signum function, page 77
$\mathbb{S}^1$ ........... unit circle, page 109
$\mathbb{S}^2$ ........... unit sphere, page 43
$L_t^2(\mathbb{S}^2)$ ...... square integrable tangential fields on $\mathbb{S}^2$, page 44
$\varepsilon$ ............. relative permittity, page 11
$\varepsilon_0$ ........... permittivity in vacuum, page 10
$|\cdot|$ ........... absolute value or euclidean norm, page 18
$\widetilde{\mathcal{T}}$ ............ middle operator for modified factorization, page 58
$A_\kappa, B_\kappa$ ...... compact operators, page 27
$B$ ............ magnetic induction, time independent, page 10
$D$ ............ electric induction, time independent, page 10
$E$ ............ electric field, time independent, page 10
$E^\infty$ ......... electric far field pattern, page 43
$E^i$ .......... incident electric field, page 12
$E^s$ .......... scattered electric field, page 12
$F_+, F_-$ ...... limit on $\Gamma$ from the exterior and interior, respectively, page 14
$G$ ............ Data–to–pattern operator, page 52
$H$ ............ magnetic field, time independent, page 10
$H(\operatorname{div}, D)$ .. functions with weak divergence, page 67
$H^\infty$ ......... magnetic far field pattern, page 43
$H^i$ .......... incident magnetic field, page 12
$H^l(\Omega)$ ....... Sobolev space of order $l$ for scalar valued functions, page 65
$H^l(\Omega, \mathbb{C}^3)$ ... Sobolev space of order $l$ for vector fields, page 30
$H^s$ .......... scattered magnetic field, page 12
$H_0(\operatorname{curl}, \Omega)$ .. vector fields with weak curl and vanishing trace, page 58
$h_n$ ........... spherical Hankel functions of the first kind, page 85
$H_{00}(\operatorname{curl}^2, \Omega)$ vector fields in $H_0(\operatorname{curl}, \Omega)$ with $\operatorname{curl} v \in H_0(\operatorname{curl}, \Omega)$, page 58

# List of Symbols

| | |
|---|---|
| $j_n$ | spherical Bessel functions, page 85 |
| $k$ | wave number, page 11 |
| $L^2(\Omega,\mathbb{C}^3)^2$ | $L^2(\Omega,\mathbb{C}^3) \times L^2(\Omega,\mathbb{C}^3)$, page 52 |
| $L^2(B,\mathbb{C}^3)$ | square integrable vector fields, page 18 |
| $L_Q^2(\Omega)$ | weighted $L^2$–space, page 57 |
| $p_\mu$ | permeability constrast, page 17 |
| $p_\varepsilon$ | permittivity contrast, page 17 |
| $P_n$ | Legendre polynomials, page 85 |
| $P_n^m$ | associated Legendre polynomials, page 84 |
| $q_\mu$ | permeability contrast, page 13 |
| $q_\varepsilon$ | permittivity contrast, page 13 |
| $T_A, T_B$ | auxiliary operators, page 27 |
| $U_n^m, V_n^m$ | vector spherical harmonics, page 86 |
| $V[G]$ | volume potential with kernel $G$, page 28 |
| $X$ | function space for modified factorization, page 58 |
| $y_n$ | spherical Neumann functions, page 85 |
| $Y_n^m$ | spherical harmonics, page 84 |
| IDE | integro–differential equation, page 22 |

# Index

achiral, 1
adjoint far field operator, 47
adjoint Herglotz operator, 53, 58, 118

Beltrami field, 91
Bessel differential equation, spherical, 85
Bessel functions, spherical, 85
Bohren's decomposition, 91

chirality, 1
circular dichroism, 2
constitutive relations, 3, 10

Data–to–pattern operator, 52, 119
Dirichlet–Neumann operator, 130, 139
Drude–Born–Fedorov, 3, 11

enantiomers, 1

far field operator, 40, 44, 118
  spherical achiral case, 99
  spherical chiral case, 102
far field pattern, 40, 43, 54
Fredholm alternative, 33, 115

fundamental solution
  in $\mathbb{R}^2$, 112
  asymptotic behavior, 40
  in $\mathbb{R}^3$, 22

Hankel functions, spherical, 85
Helmholtz equation, 22
  spherical, 84
Herglotz operator, 52, 118

injectivity
  of $\mathcal{F}$, 80
  of $\mathcal{H}$, 52
  of $\mathcal{T}$, 56, 119
integro–differential equation (IDE), 24, 26, 113
interior transmission eigenvalue, 80
inverse scattering problem, 44, 118
isotropic, 11

Jacobi–Anger expansion, 100

Lax–Milgram lemma, 27
Legendre differential equation, 85
  associated, 85

Legendre polynomials, 85
  associated, 84

Maxwell's equations, 3, 10
  for a chiral cylinder, 107
middle operator, 56, 119

Neumann functions, spherical, 85

optical activity, 2
optical rotatory dispersion, 2

radiating solution, 18, 109
radiation condition
  Silver–Müller, 18
  Sommerfeld, 109
reciprocity, 45
  principle, 45

scattering operator, 48
Stratton–Chu formulae
  exterior, 42
  interior, 41

time harmonic EM fields, 10
transmission conditions, 12
  achiral case, 14
  chiral case, 14
  cylindrical case, 108
  spherical case, 88
transmission problem, 12
  cylindrical, 109
  electric, 17
  magnetic, 13
  spherical achiral, 87
  spherical chiral, 90

variational formulation, 17, 110
vector potentials
  for Maxwell's equations, 23
  for vector Helmholtz equation, 112
vector spherical harmonics, 86
volume potential, 28

weak curl, 18
weak div, 67
weak transmission problem
  cylindrical, 111
  electric, 22
  magnetic, 21
weakly singular, 28

# Bibliography

[1] ABRAMOWITZ AND STEGUN, *Handbook of mathematical functions*, Dover Publ., 1970.

[2] R. A. ADAMS AND J. J. F. FOURNIER, *Sobolev spaces*, Academic Press, 2. ed., repr. ed., 2005.

[3] H. AMMARI AND G. BAO, *Maxwell's equations in periodic chiral structures*, Math. Nachr., 251 (2003), pp. 3–18.

[4] H. AMMARI AND J. C. NÉDÉLEC, *Time-harmonic fields in chiral media*, Meth. Verf. Math. Phys., 42 (1997), pp. 395–423.

[5] D. F. ARAGO, *Mémoire sur une modification remarquable qu' éprouvent les rayons lumineux dans leur passage à travers certains corps diaphanes et sur quelques autres nouveaux phènomènes d'optique*, Mém. Sci. Math. Phys. Inst., (1811), pp. 93–134.

[6] C. ATHANASIADIS, P. A. MARTIN, AND I. G. STRATIS, *Electromagnetic scattering by a homogeneous chiral obstacle: boundary integral equations and low-chirality approximations*, SIAM J. Appl. Math., 59 (1999), pp. 1745–1762.

[7] ——, *Electromagnetic scattering by a homogeneous chiral obstacle: Scattering relations and the far-field operator*, Math. Meth, Appl. Sci., 22 (1999), pp. 1175–1188.

[8] G. BAO AND P. LI, *Inverse medium scattering for three–dimensional time harmonic Maxwell equations*, Inverse Problems, 20 (2004), pp. L1–L7.

[9] N. BERKETIS AND C. ATHANASIADIS, *Direct and inverse scattering problems for spherical electromagnetic waves in chiral media*, ArXiv e-prints, (2008).

[10] F. CAKONI AND D. COLTON, *Qualitative Methods in Inverse Scattering Theory. An Introduction.*, Springer, Berlin, 2006.

[11] F. CAKONI, D. COLTON, AND P. MONK, *The Linear Sampling Method in Inverse Electromagnetic Scattering*, Society for Industrial and Applied Mathematics, Philadephia, PA, 2011.

[12] M. CESSENAT, *Mathematical methods in electromagnetism - linear theory and applications*, vol. 41 of Series on advances in mathematics for applied sciences, World Scientific, 1996.

[13] D. COLTON, H. HADDAR, AND M. PIANA, *The linear sampling method in inverse electromagnetic scattering theory*, Inverse Problems, 19 (2003), pp. S105–S137.

[14] D. L. COLTON AND R. KRESS, *Integral equation methods in scattering theory*, Krieger, repr. ed., with corr. ed., 1992.

[15] ———, *Inverse acoustic and electromagnetic scattering theory*, Springer, 2nd ed., 1998.

[16] F. CRAIG AND BOHREN, *Light scattering by an optically active sphere*, Chemical Physics Letters, 29 (1974), pp. 458 – 462.

[17] O. DORN, H. BERTETE-AGUIRRE, J. G. BERRYMAN, AND G. C. PAPANICOLAOU, *A nonlinear inversion method for 3d electromagnetic imaging using adjoint fields*, Inverse Problems, 15 (1999), pp. 1523–1558.

[18] P. DRUDE, *Lehrbuch der Optik*, Hirzel, Leipzig, 2., erw. aufl. ed., 1906.

[19] J. HADAMARD, *Lectures on the Cauchy Problem in Linear Partial Differential Equations*, Yale University Press, New Haven, 1923.

[20] A. KIRSCH, *The factorization method for Maxwell's equations*, Inverse Problems, 20 (2004), pp. S117–S134.

[21] ———, *An integral equation approach and the interior transmission problem for Maxwell's equations*, Inverse Problems and Imaging, 1 (2007), pp. 159–180.

[22] ———, *An integral equation for Maxwell's equations in a layered medium with an application to the Factorization method*, J. Integral Equations Appl., 19 (2007), pp. 333–358.

[23] ———, *An integral equation for the scattering problem for an anisotropic medium and the Factorization method*, in 8th int. workshop on mathematical methods in scattering theory and biomedical engineering, 2007.

[24] A. KIRSCH AND N. GRINBERG, *The Factorization Method for Inverse Problems*, Oxford Lecture Series in Mathematics and its Applications 36, Oxford University Press, 2008.

[25] R. KRESS, *Linear integral equations*, Springer, 1989.

[26] A. LAKHTAKIA, V. K. VARADAN, AND V. V. VARADAN, *Time harmonic electromagnetic fields in chiral media*, vol. 335 of Lecture Notes in Physics, Springer, 1989.

[27] A. LECHLEITER, *The Factorization method is independent of transmission eigenvalues*, Inverse Problems and Imaging, 3 (2009), pp. 123–138.

[28] A. LECHLEITER AND S. RITTERBUSCH, *Scattering of acoustic waves from rough layers*, in progress, (2009).

[29] K. LINDMAN, *Über die durch ein isotropes System von spiralförmigen Resonatoren erzeugte Rotationspolarisation der elektromagnetischen Wellen*, Ann. Physik, 63 (1920), p. 621.

[30] W. McLEAN, *Strongly Elliptic Systems and Boundary Integral Operators*, Cambridge University Press, Cambridge, UK, 2000.

[31] P. MONK, *Finite Element Methods for Maxwell's Equations*, Oxford Science Publications, Oxford, 2003.

[32] L. PASTEUR, *Sur les relations qui peuvent exister entre la forme christalline, la composante chimique et le sens de la polarisation rotataire*, Ann. Chimie et Physique, 24 (1848), pp. 442–459.

[33] R. POTTHAST AND M. SINI, *The no response test for the reconstruction of polyhedral objects in electromagnetics*, J. Comp. Appl. Math., 234 (2010), pp. 1739 – 1746.

[34] C. RAMANANJAONA, M. LAMBERT, AND D. LESSELIER, *Shape inversion from tm and te real data by controlled evolution of level sets*, Inverse Problems, 17 (2001), pp. 1585–1595.

[35] R. G. ROJAS, *Intgeral equations for the scattering by a three dimensional inhomogeneous chiral body*, J. Electromagn. Waves Appl., 6 (1992), pp. 733–750.

[36] ——, *Intgeral equations for EM scattering by homogeneous/inhomogeneous two-dimensional chiral bodies*, IEE Proc.-Microw. Antennas Propag., 141 (1994), pp. 385–392.

[37] K. SANDFORT, *The factorization method for inverse scattering from periodic inhomogeneous media*, PhD thesis, Department of Mathematics, KIT, 2010.

[38] M. VACATELLO AND P. J. FLORY, *Helical conformations of isotactic poly (methyl methacrylate). Energies computed with bond angle relaxation*, Polym. Commun., 25 (1984), p. 258.

[39] C. WEBER, *A local compactness theorem for Maxwell's equations*, Math. Methods Appl. Sci., 2 (1980), pp. 12–25.

Printed by Books on Demand GmbH, Norderstedt / Germany